Praise for *Learning Spark*, Second Edition

This book offers a structured approach to learning Apache Spark, covering new developments in the project. It is a great way for Spark developers to get started with big data.

—Reynold Xin, Databricks Chief Architect and Cofounder and Apache Spark PMC Member

For data scientists and data engineers looking to learn Apache Spark and how to build scalable and reliable big data applications, this book is an essential guide!

—Ben Lorica, Databricks Chief Data Scientist, Past Program Chair O'Reilly Strata Conferences, Program Chair for Spark + AI Summit

SECOND EDITION

Learning Spark

Lightning-Fast Data Analytics

Jules S. Damji, Brooke Wenig,
Tathagata Das, and Denny Lee

Beijing · Boston · Farnham · Sebastopol · Tokyo

Learning Spark

by Jules S. Damji, Brooke Wenig, Tathagata Das, and Denny Lee

Printed in the United States of America.

Published by O'Reilly Media, Inc., 1005 Gravenstein Highway North, Sebastopol, CA 95472.

O'Reilly books may be purchased for educational, business, or sales promotional use. Online editions are also available for most titles (*http://oreilly.com*). For more information, contact our corporate/institutional sales department: 800-998-9938 or *corporate@oreilly.com*.

Acquisitions Editor: Jonathan Hassell	**Indexer:** Potomac Indexing, LLC
Development Editor: Michele Cronin	**Interior Designer:** David Futato
Production Editor: Deborah Baker	**Cover Designer:** Karen Montgomery
Copyeditor: Rachel Head	**Illustrator:** Rebecca Demarest
Proofreader: Penelope Perkins	

January 2015:   First Edition
July 2020:   Second Edition

Revision History for the Second Edition

2020-06-24:   First Release
2020-08-03:   Second Release
2020-11-20:   Third Release
2022-09-30:   Fourth Release

See *http://oreilly.com/catalog/errata.csp?isbn=9781492050049* for release details.

978-1-492-05004-9

[LSI]

Table of Contents

Foreword

Apache Spark has evolved significantly since I first started the project at UC Berkeley in 2009. After moving to the Apache Software Foundation, the open source project has had over 1,400 contributors from hundreds of companies, and the global Spark meetup group (*https://oreil.ly/SB8S-*) has grown to over half a million members. Spark's user base has also become highly diverse, encompassing Python, R, SQL, and JVM developers, with use cases ranging from data science to business intelligence to data engineering. I have been working closely with the Apache Spark community to help continue its development, and I am thrilled to see the progress thus far.

The release of Spark 3.0 marks an important milestone for the project and has sparked the need for updated learning material. The idea of a second edition of *Learning Spark* has come up many times—and it was overdue. Even though I coauthored both *Learning Spark* and *Spark: The Definitive Guide* (both O'Reilly), it was time for me to let the next generation of Spark contributors pick up the narrative. I'm delighted that four experienced practitioners and developers, who have been working closely with Apache Spark from its early days, have teamed up to write this second edition of the book, incorporating the most recent APIs and best practices for Spark developers in a clear and informative guide.

The authors' approach to this edition is highly conducive to hands-on learning. The key concepts in Spark and distributed big data processing have been distilled into easy-to-follow chapters. Through the book's illustrative code examples, developers can build confidence using Spark and gain a greater understanding of its Structured APIs and how to leverage them. I hope that this second edition of *Learning Spark* will guide you on your large-scale data processing journey, whatever problems you wish to tackle using Spark.

— Matei Zaharia, Chief Technologist,
Cofounder of Databricks, Asst. Professor at Stanford,
and original creator of Apache Spark

Preface

We welcome you to the second edition of *Learning Spark*. It's been five years since the first edition was published in 2015, originally authored by Holden Karau, Andy Konwinski, Patrick Wendell, and Matei Zaharia. This new edition has been updated to reflect Apache Spark's evolution through Spark 2.x and Spark 3.0, including its expanded ecosystem of built-in and external data sources, machine learning, and streaming technologies with which Spark is tightly integrated.

Over the years since its first 1.x release, Spark has become the de facto big data unified processing engine. Along the way, it has extended its scope to include support for various analytic workloads. Our intent is to capture and curate this evolution for readers, showing not only how you can use Spark but how it fits into the new era of big data and machine learning. Hence, we have designed each chapter to build progressively on the foundations laid by the previous chapters, ensuring that the content is suited for our intended audience.

Who This Book Is For

Most developers who grapple with big data are data engineers, data scientists, or machine learning engineers. This book is aimed at those professionals who are looking to use Spark to scale their applications to handle massive amounts of data.

In particular, data engineers will learn how to use Spark's Structured APIs to perform complex data exploration and analysis on both batch and streaming data; use Spark SQL for interactive queries; use Spark's built-in and external data sources to read, refine, and write data in different file formats as part of their extract, transform, and load (ETL) tasks; and build reliable data lakes with Spark and the open source Delta Lake table format.

For data scientists and machine learning engineers, Spark's MLlib library offers many common algorithms to build distributed machine learning models. We will cover how to build pipelines with MLlib, best practices for distributed machine learning, how to use Spark to scale single-node models, and how to manage and deploy these models using the open source library MLflow.

While the book is focused on learning Spark as an analytical engine for diverse work-loads, we will not cover all of the languages that Spark supports. Most of the examples in the chapters are written in Scala, Python, and SQL. Where necessary, we have infused a bit of Java. For those interested in learning Spark with R, we recommend Javier Luraschi, Kevin Kuo, and Edgar Ruiz's *Mastering Spark with R* (O'Reilly).

Finally, because Spark is a distributed engine, building an understanding of Spark application concepts is critical. We will guide you through how your Spark applica-tion interacts with Spark's distributed components and how execution is decomposed into parallel tasks on a cluster. We will also cover which deployment modes are sup-ported and in what environments.

While there are many topics we have chosen to cover, there are a few that we have opted to not focus on. These include the older low-level Resilient Distributed Dataset (RDD) APIs and GraphX, Spark's API for graphs and graph-parallel computation. Nor have we covered advanced topics such as how to extend Spark's Catalyst opti-mizer to implement your own operations, how to implement your own catalog, or how to write your own DataSource V2 data sinks and sources. Though part of Spark, these are beyond the scope of your first book on learning Spark.

Instead, we have focused and organized the book around Spark's Structured APIs, across all its components, and how you can use Spark to process structured data at scale to perform your data engineering or data science tasks.

How the Book Is Organized

We organized the book in a way that leads you from chapter to chapter by introduc-ing concepts, demonstrating these concepts via example code snippets, and providing full code examples or notebooks in the book's GitHub repo (*https://github.com/data bricks/LearningSparkV2*).

Chapter 1, *"Introduction to Apache Spark: A Unified Analytics Engine"*
> Introduces you to the evolution of big data and provides a high-level overview of Apache Spark and its application to big data.

Chapter 2, *"Downloading Apache Spark and Getting Started"*
> Walks you through downloading and setting up Apache Spark on your local machine.

Chapter 3, *"Apache Spark's Structured APIs"* through Chapter 6, *"Spark SQL and Data-sets"*
> These chapters focus on using the DataFrame and Dataset Structured APIs to ingest data from built-in and external data sources, apply built-in and custom

functions, and utilize Spark SQL. These chapters comprise the foundation for later chapters, incorporating all the latest Spark 3.0 changes where appropriate.

Chapter 7, "Optimizing and Tuning Spark Applications"
Provides you with best practices for tuning, optimizing, debugging, and inspecting your Spark applications through the Spark UI, as well as details on the configurations you can tune to increase performance.

Chapter 8, "Structured Streaming"
Guides you through the evolution of the Spark Streaming engine and the Structured Streaming programming model. It examines the anatomy of a typical streaming query and discusses the different ways to transform streaming data—stateful aggregations, stream joins, and arbitrary stateful aggregation—while providing guidance on how to design performant streaming queries.

Chapter 9, "Building Reliable Data Lakes with Apache Spark"
Surveys three open source table format storage solutions, as part of the Spark ecosystem, that employ Apache Spark to build reliable data lakes with transactional guarantees. Due to Delta Lake's tight integration with Spark for both batch and streaming workloads, we focus on that solution and explore how it facilitates a new paradigm in data management, the lakehouse.

Chapter 10, "Machine Learning with MLlib"
Introduces MLlib, the distributed machine learning library for Spark, and walks you through an end-to-end example of how to build a machine learning pipeline, including topics such as feature engineering, hyperparameter tuning, evaluation metrics, and saving and loading models.

Chapter 11, "Managing, Deploying, and Scaling Machine Learning Pipelines with Apache Spark"
Covers how to track and manage your MLlib models with MLflow, compares and contrasts different model deployment options, and explores how to leverage Spark for non-MLlib models for distributed model inference, feature engineering, and/or hyperparameter tuning.

Chapter 12, "Epilogue: Apache Spark 3.0"
The epilogue highlights notable features and changes in Spark 3.0. While the full range of enhancements and features is too extensive to fit in a single chapter, we highlight the major changes you should be aware of and recommend you check the release notes when Spark 3.0 is officially released.

Throughout these chapters, we have incorporated or noted Spark 3.0 features where needed and tested all the code examples and notebooks against Spark 3.0.0-preview2.

How to Use the Code Examples

The code examples in the book range from brief snippets to complete Spark applications and end-to-end notebooks, in Scala, Python, SQL, and, where necessary, Java.

While some short code snippets in a chapter are self-contained and can be copied and pasted to run in a Spark shell (`pyspark` or `spark-shell`), others are fragments from standalone Spark applications or end-to-end notebooks. To run standalone Spark applications in Scala, Python, or Java, read the instructions in the respective chapter's README files in this book's GitHub repo (*https://github.com/databricks/Learning SparkV2*).

As for the notebooks, to run these you will need to register for a free Databricks Community Edition (*https://community.cloud.databricks.com/*) account. We detail how to import the notebooks and create a cluster using Spark 3.0 in the README (*https://github.com/databricks/LearningSparkV2/tree/master/notebooks*).

Software and Configuration Used

Most of the code in this book and the accompanying notebooks were written in and tested against Apache Spark 3.0.0-preview2, which was available to us at the time we were writing the final chapters.

By the time this book is published, Apache Spark 3.0 will have been released and be available to the community for general use. We recommend that you download (*https://oreil.ly/WFX48*) and use the official release with the following configurations for your operating system:

- Apache Spark 3.0 (prebuilt for Apache Hadoop 2.7)
- Java Development Kit (JDK) 1.8.0

If you intend to use only Python, then you can simply run `pip install pyspark`.

Conventions Used in This Book

The following typographical conventions are used in this book:

Italic
> Indicates new terms, URLs, email addresses, filenames, and file extensions.

`Constant width`
> Used for program listings, as well as within paragraphs to refer to program elements such as variable or function names, databases, data types, environment variables, statements, and keywords.

Constant width bold

Shows commands or other text that should be typed literally by the user.

Constant width italic

Shows text that should be replaced with user-supplied values or by values determined by context.

 This element signifies a general note.

Using Code Examples

If you have a technical question or a problem using the code examples, please send an email to *bookquestions@oreilly.com*.

This book is here to help you get your job done. In general, if example code is offered with this book, you may use it in your programs and documentation. You do not need to contact us for permission unless you're reproducing a significant portion of the code. For example, writing a program that uses several chunks of code from this book does not require permission. Selling or distributing examples from O'Reilly books does require permission. Answering a question by citing this book and quoting example code does not require permission. Incorporating a significant amount of example code from this book into your product's documentation does require permission.

We appreciate, but generally do not require, attribution. An attribution usually includes the title, author, publisher, and ISBN. For example: "*Learning Spark*, 2nd Edition, by Jules S. Damji, Brooke Wenig, Tathagata Das, and Denny Lee. Copyright 2020 Databricks, Inc., 978-1-492-05004-9."

If you feel your use of code examples falls outside fair use or the permission given above, feel free to contact us at *permissions@oreilly.com*.

O'Reilly Online Learning

 For more than 40 years, *O'Reilly Media* has provided technology and business training, knowledge, and insight to help companies succeed.

Our unique network of experts and innovators share their knowledge and expertise through books, articles, and our online learning platform. O'Reilly's online learning platform gives you on-demand access to live training courses, in-depth learning paths, interactive coding environments, and a vast collection of text and video from O'Reilly and 200+ other publishers. For more information, visit *http://oreilly.com*.

How to Contact Us

Please address comments and questions concerning this book to the publisher:

O'Reilly Media, Inc.
1005 Gravenstein Highway North
Sebastopol, CA 95472
800-998-9938 (in the United States or Canada)
707-829-0515 (international or local)
707-829-0104 (fax)

Visit our web page for this book, where we list errata, examples, and any additional information, at *https://oreil.ly/LearningSpark2*.

Email *bookquestions@oreilly.com* to comment or ask technical questions about this book.

For news and information about our books and courses, visit *http://oreilly.com*.

Find us on Facebook: *http://facebook.com/oreilly*

Follow us on Twitter: *http://twitter.com/oreillymedia*

Watch us on YouTube: *http://www.youtube.com/oreillymedia*

Acknowledgments

This project was truly a team effort involving many people, and without their support and feedback we would not have been able to finish this book, especially in today's unprecedented COVID-19 times.

First and foremost, we want to thank our employer, Databricks, for supporting us and allocating us dedicated time as part of our jobs to finish this book. In particular, we want to thank Matei Zaharia, Reynold Xin, Ali Ghodsi, Ryan Boyd, and Rick Schultz for encouraging us to write the second edition.

Second, we would like to thank our technical reviewers: Adam Breindel, Amir Issaei, Jacek Laskowski, Sean Owen, and Vishwanath Subramanian. Their diligent and constructive feedback, informed by their technical expertise in the community and industry point of view, made this book what it is: a valuable resource to learn Spark.

Besides the formal book reviewers, we received invaluable feedback from others knowledgeable about specific topics and sections of the chapters, and we want to acknowledge their contributions. Many thanks to: Conor Murphy, Hyukjin Kwon, Maryann Xue, Niall Turbitt, Wenchen Fan, Xiao Li, and Yuanjian Li.

Finally, we would like to thank our colleagues at Databricks (for their tolerance of us missing or neglecting project deadlines), our families and loved ones (for their patience and empathy as we wrote in the early light of day or late into the night on weekdays and weekends), and the entire open source Spark community. Without their continued contributions, Spark would not be where it is today—and we authors would not have had much to write about.

Thank you all!

CHAPTER 1

Introduction to Apache Spark: A Unified Analytics Engine

This chapter lays out the origins of Apache Spark and its underlying philosophy. It also surveys the main components of the project and its distributed architecture. If you are familiar with Spark's history and the high-level concepts, you can skip this chapter.

The Genesis of Spark

In this section, we'll chart the course of Apache Spark's short evolution: its genesis, inspiration, and adoption in the community as a de facto big data unified processing engine.

Big Data and Distributed Computing at Google

When we think of scale, we can't help but think of the ability of Google's search engine to index and search the world's data on the internet at lightning speed. The name Google is synonymous with scale. In fact, Google is a deliberate misspelling of the mathematical term *googol*: that's 1 plus 100 zeros!

Neither traditional storage systems such as relational database management systems (RDBMSs) nor imperative ways of programming were able to handle the scale at which Google wanted to build and search the internet's indexed documents. The resulting need for new approaches led to the creation of the *Google File System* (GFS) (*https://oreil.ly/-6H9D*), *MapReduce* (MR) (*https://oreil.ly/08zaO*), and *Bigtable* (*https://oreil.ly/KfS8C*).

While GFS provided a fault-tolerant and distributed filesystem across many commodity hardware servers in a cluster farm, Bigtable offered scalable storage of

1

structured data across GFS. MR introduced a new parallel programming paradigm, based on functional programming, for large-scale processing of data distributed over GFS and Bigtable.

In essence, your MR applications interact with the MapReduce system (*https://oreil.ly/T0f8r*) that sends computation code (map and reduce functions) to where the data resides, favoring data locality and cluster rack affinity rather than bringing data to your application.

The workers in the cluster aggregate and reduce the intermediate computations and produce a final appended output from the reduce function, which is then written to a distributed storage where it is accessible to your application. This approach significantly reduces network traffic and keeps most of the input/output (I/O) local to disk rather than distributing it over the network.

Most of the work Google did was proprietary, but the ideas expressed in the aforementioned three papers spurred innovative ideas elsewhere in the open source community—especially at Yahoo!, which was dealing with similar big data challenges of scale for its search engine.

Hadoop at Yahoo!

The computational challenges and solutions expressed in Google's GFS paper provided a blueprint for the Hadoop File System (HDFS) (*https://oreil.ly/JfsBd*), including the MapReduce implementation as a framework for distributed computing. Donated to the Apache Software Foundation (ASF) (*https://www.apache.org/*), a vendor-neutral non-profit organization, in April 2006, it became part of the Apache Hadoop (*https://oreil.ly/twL6R*) framework of related modules: Hadoop Common, MapReduce, HDFS, and Apache Hadoop YARN.

Although Apache Hadoop had garnered widespread adoption outside Yahoo!, inspiring a large open source community of contributors and two open source–based commercial companies (Cloudera and Hortonworks, now merged), the MapReduce framework on HDFS had a few shortcomings.

First, it was hard to manage and administer, with cumbersome operational complexity. Second, its general batch-processing MapReduce API was verbose and required a lot of boilerplate setup code, with brittle fault tolerance. Third, with large batches of data jobs with many pairs of MR tasks, each pair's intermediate computed result is written to the local disk for the subsequent stage of its operation (see Figure 1-1). This repeated performance of disk I/O took its toll: large MR jobs could run for hours on end, or even days.

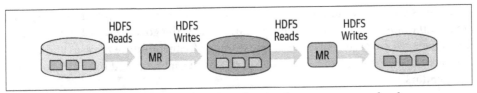

Figure 1-1. Intermittent iteration of reads and writes between map and reduce computations

And finally, even though Hadoop MR was conducive to large-scale jobs for general batch processing, it fell short for combining other workloads such as machine learning, streaming, or interactive SQL-like queries.

To handle these new workloads, engineers developed bespoke systems (Apache Hive, Apache Storm, Apache Impala, Apache Giraph, Apache Drill, Apache Mahout, etc.), each with their own APIs and cluster configurations, further adding to the operational complexity of Hadoop and the steep learning curve for developers.

The question then became (bearing in mind Alan Kay's adage, "Simple things should be simple, complex things should be possible"), was there a way to make Hadoop and MR simpler and faster?

Spark's Early Years at AMPLab

Researchers at UC Berkeley who had previously worked on Hadoop MapReduce took on this challenge with a project they called *Spark*. They acknowledged that MR was inefficient (or intractable) for interactive or iterative computing jobs and a complex framework to learn, so from the onset they embraced the idea of making Spark simpler, faster, and easier. This endeavor started in 2009 at the RAD Lab, which later became the AMPLab (and now is known as the RISELab).

Early papers (*https://oreil.ly/RFY2w*) published on Spark demonstrated that it was 10 to 20 times faster than Hadoop MapReduce for certain jobs. Today, it's many orders of magnitude faster (*https://spark.apache.org*). The central thrust of the Spark project was to bring in ideas borrowed from Hadoop MapReduce, but to enhance the system: make it highly fault tolerant and embarrassingly parallel, support in-memory storage for intermediate results between iterative and interactive map and reduce computations, offer easy and composable APIs in multiple languages as a programming model, and support other workloads in a unified manner. We'll come back to this idea of unification shortly, as it's an important theme in Spark.

By 2013 Spark had gained widespread use, and some of its original creators and researchers—Matei Zaharia, Ali Ghodsi, Reynold Xin, Patrick Wendell, Ion Stoica, and Andy Konwinski—donated the Spark project to the ASF and formed a company called Databricks.

Databricks and the community of open source developers worked to release Apache Spark 1.0 (*https://oreil.ly/Pq11v*) in May 2014, under the governance of the ASF. This first major release established the momentum for frequent future releases and contributions of notable features to Apache Spark from Databricks and over 100 commercial vendors.

What Is Apache Spark?

Apache Spark (*https://spark.apache.org*) is a unified engine designed for large-scale distributed data processing, on premises in data centers or in the cloud.

Spark provides in-memory storage for intermediate computations, making it much faster than Hadoop MapReduce. It incorporates libraries with composable APIs for machine learning (MLlib), SQL for interactive queries (Spark SQL), stream processing (Structured Streaming) for interacting with real-time data, and graph processing (GraphX).

Spark's design philosophy centers around four key characteristics:

- Speed
- Ease of use
- Modularity
- Extensibility

Let's take a look at what this means for the framework.

Speed

Spark has pursued the goal of speed in several ways. First, its internal implementation benefits immensely from the hardware industry's recent huge strides in improving the price and performance of CPUs and memory. Today's commodity servers come cheap, with hundreds of gigabytes of memory, multiple cores, and the underlying Unix-based operating system taking advantage of efficient multithreading and parallel processing. The framework is optimized to take advantage of all of these factors.

Second, Spark builds its query computations as a directed acyclic graph (DAG); its DAG scheduler and query optimizer construct an efficient computational graph that can usually be decomposed into tasks that are executed in parallel across workers on the cluster. And third, its physical execution engine, Tungsten, uses whole-stage code generation to generate compact code for execution (we will cover SQL optimization and whole-stage code generation in Chapter 3).

With all the intermediate results retained in memory and its limited disk I/O, this gives it a huge performance boost.

Ease of Use

Spark achieves simplicity by providing a fundamental abstraction of a simple logical data structure called a Resilient Distributed Dataset (RDD) upon which all other higher-level structured data abstractions, such as DataFrames and Datasets, are constructed. By providing a set of *transformations* and *actions* as *operations*, Spark offers a simple programming model that you can use to build big data applications in familiar languages.

Modularity

Spark operations can be applied across many types of workloads and expressed in any of the supported programming languages: Scala, Java, Python, SQL, and R. Spark offers unified libraries with well-documented APIs that include the following modules as core components: Spark SQL, Spark Structured Streaming, Spark MLlib, and GraphX, combining all the workloads running under one engine. We'll take a closer look at all of these in the next section.

You can write a single Spark application that can do it all—no need for distinct engines for disparate workloads, no need to learn separate APIs. With Spark, you get a unified processing engine for your workloads.

Extensibility

Spark focuses on its fast, parallel computation engine rather than on storage. Unlike Apache Hadoop, which included both storage and compute, Spark decouples the two. That means you can use Spark to read data stored in myriad sources—Apache Hadoop, Apache Cassandra, Apache HBase, MongoDB, Apache Hive, RDBMSs, and more—and process it all in memory. Spark's `DataFrameReaders` and `DataFrame Writers` can also be extended to read data from other sources, such as Apache Kafka, Kinesis, Azure Storage, and Amazon S3, into its logical data abstraction, on which it can operate.

The community of Spark developers maintains a list of third-party Spark packages (*https://oreil.ly/2tIVP*) as part of the growing ecosystem (see Figure 1-2). This rich ecosystem of packages includes Spark connectors for a variety of external data sources, performance monitors, and more.

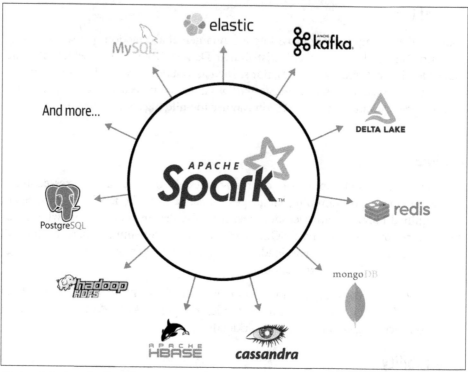

Figure 1-2. Apache Spark's ecosystem of connectors

Unified Analytics

While the notion of unification is not unique to Spark, it is a core component of its design philosophy and evolution. In November 2016, the Association for Computing Machinery (ACM) recognized Apache Spark and conferred upon its original creators the prestigious ACM Award for their paper (*https://oreil.ly/eak-T*) describing Apache Spark as a "Unified Engine for Big Data Processing." The award-winning paper notes that Spark replaces all the separate batch processing, graph, stream, and query engines like Storm, Impala, Dremel, Pregel, etc. with a unified stack of components that addresses diverse workloads under a single distributed fast engine.

Apache Spark Components as a Unified Stack

As shown in Figure 1-3, Spark offers four distinct components as libraries for diverse workloads: Spark SQL, Spark MLlib, Spark Structured Streaming, and GraphX. Each of these components is separate from Spark's core fault-tolerant engine, in that you use APIs to write your Spark application and Spark converts this into a DAG that is executed by the core engine. So whether you write your Spark code using the provided Structured APIs (which we will cover in Chapter 3) in Java, R, Scala, SQL, or

Python, the underlying code is decomposed into highly compact bytecode that is executed in the workers' JVMs across the cluster.

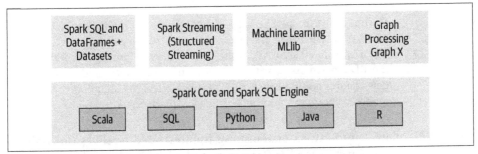

Figure 1-3. Apache Spark components and API stack

Let's look at each of these components in more detail.

Spark SQL

This module works well with structured data. You can read data stored in an RDBMS table or from file formats with structured data (CSV, text, JSON, Avro, ORC, Parquet, etc.) and then construct permanent or temporary tables in Spark. Also, when using Spark's Structured APIs in Java, Python, Scala, or R, you can combine SQL-like queries to query the data just read into a Spark DataFrame. To date, Spark SQL is ANSI SQL:2003-compliant (*https://oreil.ly/pJq1C*) and it also functions as a pure SQL engine.

For example, in this Scala code snippet, you can read from a JSON file stored on Amazon S3, create a temporary table, and issue a SQL-like query on the results read into memory as a Spark DataFrame:

```
// In Scala
// Read data off Amazon S3 bucket into a Spark DataFrame
spark.read.json("s3://apache_spark/data/committers.json")
  .createOrReplaceTempView("committers")
// Issue a SQL query and return the result as a Spark DataFrame
val results = spark.sql("""SELECT name, org, module, release, num_commits
    FROM committers WHERE module = 'mllib' AND num_commits > 10
    ORDER BY num_commits DESC""")
```

You can write similar code snippets in Python, R, or Java, and the generated bytecode will be identical, resulting in the same performance.

Spark MLlib

Spark comes with a library containing common machine learning (ML) algorithms called MLlib. Since Spark's first release, the performance of this library component has improved significantly because of Spark 2.x's underlying engine enhancements.

MLlib provides many popular machine learning algorithms built atop high-level DataFrame-based APIs to build models.

 Starting with Apache Spark 1.6, the MLlib project (*https://oreil.ly/cyc1c*) is split between two packages: `spark.mllib` and `spark.ml`. The DataFrame-based API is the latter while the former contains the RDD-based APIs, which are now in maintenance mode. All new features go into `spark.ml`. This book refers to "MLlib" as the umbrella library for machine learning in Apache Spark.

These APIs allow you to extract or transform features, build pipelines (for training and evaluating), and persist models (for saving and reloading them) during deployment. Additional utilities include the use of common linear algebra operations and statistics. MLlib includes other low-level ML primitives, including a generic gradient descent optimization. The following Python code snippet encapsulates the basic operations a data scientist may do when building a model (more extensive examples will be discussed in Chapters 10 and 11):

```
# In Python
from pyspark.ml.classification import LogisticRegression
...
training = spark.read.csv("s3://...")
test = spark.read.csv("s3://...")

# Load training data
lr = LogisticRegression(maxIter=10, regParam=0.3, elasticNetParam=0.8)

# Fit the model
lrModel = lr.fit(training)

# Predict
lrModel.transform(test)
...
```

Spark Structured Streaming

Apache Spark 2.0 introduced an experimental Continuous Streaming model (*https://oreil.ly/YJSEq*) and Structured Streaming APIs (*https://oreil.ly/NYYsJ*), built atop the Spark SQL engine and DataFrame-based APIs. By Spark 2.2, Structured Streaming was generally available, meaning that developers could use it in their production environments.

Necessary for big data developers to combine and react in real time to both static data and streaming data from engines like Apache Kafka and other streaming sources, the new model views a stream as a continually growing table, with new rows of data appended at the end. Developers can merely treat this as a structured table and issue queries against it as they would a static table.

Underneath the Structured Streaming model, the Spark SQL core engine handles all aspects of fault tolerance and late-data semantics, allowing developers to focus on writing streaming applications with relative ease. This new model obviated the old DStreams model in Spark's 1.x series, which we will discuss in more detail in Chapter 8. Furthermore, Spark 2.x and Spark 3.0 extended the range of streaming data sources to include Apache Kafka, Kinesis, and HDFS-based or cloud storage.

The following code snippet shows the typical anatomy of a Structured Streaming application. It reads from a localhost socket and writes the word count results to an Apache Kafka topic:

```
# In Python
# Read a stream from a local host
from pyspark.sql.functions import explode, split
lines = (spark
  .readStream
  .format("socket")
  .option("host", "localhost")
  .option("port", 9999)
  .load())

# Perform transformation
# Split the lines into words
words = lines.select(explode(split(lines.value, " ")).alias("word"))

# Generate running word count
word_counts = words.groupBy("word").count()

# Write out to the stream to Kafka
query = (word_counts
  .writeStream
  .format("kafka")
  .option("topic", "output"))
```

GraphX

As the name suggests, GraphX is a library for manipulating graphs (e.g., social network graphs, routes and connection points, or network topology graphs) and performing graph-parallel computations. It offers the standard graph algorithms for analysis, connections, and traversals, contributed by users in the community: the available algorithms include PageRank, Connected Components, and Triangle Counting.[1]

This code snippet shows a simple example of how to join two graphs using the GraphX APIs:

---

1 Contributed to the community by Databricks as an open source project, GraphFrames (*https://oreil.ly/_JGxi*) is a general graph processing library that is similar to Apache Spark's GraphX but uses DataFrame-based APIs.

```
// In Scala
val graph = Graph(vertices, edges)
messages = spark.textFile("hdfs://...")
val graph2 = graph.joinVertices(messages) {
  (id, vertex, msg) => ...
}
```

Apache Spark's Distributed Execution

If you have read this far, you already know that Spark is a distributed data processing engine with its components working collaboratively on a cluster of machines. Before we explore programming with Spark in the following chapters of this book, you need to understand how all the components of Spark's distributed architecture work together and communicate, and what deployment modes are available.

Let's start by looking at each of the individual components shown in Figure 1-4 and how they fit into the architecture. At a high level in the Spark architecture, a Spark application consists of a driver program that is responsible for orchestrating parallel operations on the Spark cluster. The driver accesses the distributed components in the cluster—the Spark executors and cluster manager—through a SparkSession.

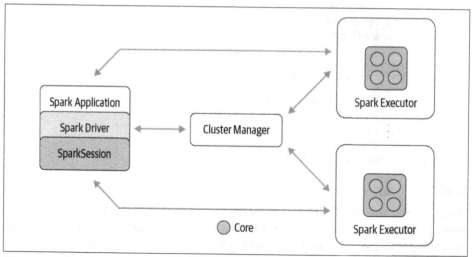

Figure 1-4. Apache Spark components and architecture

Spark driver

As the part of the Spark application responsible for instantiating a SparkSession, the Spark driver has multiple roles: it communicates with the cluster manager; it requests resources (CPU, memory, etc.) from the cluster manager for Spark's executors (JVMs); and it transforms all the Spark operations into DAG computations, schedules them, and distributes their execution as tasks across the Spark executors. Once the resources are allocated, it communicates directly with the executors.

SparkSession

In Spark 2.0, the `SparkSession` became a unified conduit to all Spark operations and data. Not only did it subsume previous entry points to Spark (*https://oreil.ly/Ap0Pq*) like the `SparkContext`, `SQLContext`, `HiveContext`, `SparkConf`, and `StreamingCon text`, but it also made working with Spark simpler and easier.

 Although in Spark 2.x the `SparkSession` subsumes all other contexts, you can still access the individual contexts and their respective methods. In this way, the community maintained backward compatibility. That is, your old 1.x code with `SparkContext` or `SQLContext` will still work.

Through this one conduit, you can create JVM runtime parameters, define Data-Frames and Datasets, read from data sources, access catalog metadata, and issue Spark SQL queries. `SparkSession` provides a single unified entry point to all of Spark's functionality.

In a standalone Spark application, you can create a `SparkSession` using one of the high-level APIs in the programming language of your choice. In the Spark shell (more on this in the next chapter) the `SparkSession` is created for you, and you can access it via a global variable called `spark` or `sc`.

Whereas in Spark 1.x you would have had to create individual contexts (for streaming, SQL, etc.), introducing extra boilerplate code, in a Spark 2.x application you can create a `SparkSession` per JVM and use it to perform a number of Spark operations.

Let's take a look at an example:

```scala
// In Scala
import org.apache.spark.sql.SparkSession

// Build SparkSession
val spark = SparkSession
  .builder
  .appName("LearnSpark")
  .config("spark.sql.shuffle.partitions", 6)
  .getOrCreate()
...
// Use the session to read JSON
val people = spark.read.json("...")
...
// Use the session to issue a SQL query
val resultsDF = spark.sql("SELECT city, pop, state, zip FROM table_name")
```

Cluster manager

The cluster manager is responsible for managing and allocating resources for the cluster of nodes on which your Spark application runs. Currently, Spark supports four cluster managers: the built-in standalone cluster manager, Apache Hadoop YARN, Apache Mesos, and Kubernetes.

Spark executor

A Spark executor runs on each worker node in the cluster. The executors communicate with the driver program and are responsible for executing tasks on the workers. In most deployments modes, only a single executor runs per node.

Deployment modes

An attractive feature of Spark is its support for myriad deployment modes, enabling Spark to run in different configurations and environments. Because the cluster manager is agnostic to where it runs (as long as it can manage Spark's executors and fulfill resource requests), Spark can be deployed in some of the most popular environments—such as Apache Hadoop YARN and Kubernetes—and can operate in different modes. Table 1-1 summarizes the available deployment modes.

Table 1-1. Cheat sheet for Spark deployment modes

Mode	Spark driver	Spark executor	Cluster manager
Local	Runs on a single JVM, like a laptop or single node	Runs on the same JVM as the driver	Runs on the same host
Standalone	Can run on any node in the cluster	Each node in the cluster will launch its own executor JVM	Can be allocated arbitrarily to any host in the cluster
YARN (client)	Runs on a client, not part of the cluster	YARN's NodeManager's container	YARN's Resource Manager works with YARN's Application Master to allocate the containers on NodeManagers for executors
YARN (cluster)	Runs with the YARN Application Master	Same as YARN client mode	Same as YARN client mode
Kubernetes	Runs in a Kubernetes pod	Each worker runs within its own pod	Kubernetes Master

Distributed data and partitions

Actual physical data is distributed across storage as partitions residing in either HDFS or cloud storage (see Figure 1-5). While the data is distributed as partitions across the physical cluster, Spark treats each partition as a high-level logical data abstraction—as a DataFrame in memory. Though this is not always possible, each Spark executor is preferably allocated a task that requires it to read the partition closest to it in the network, observing data locality.

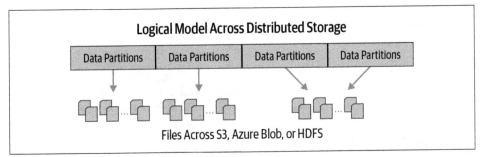

Logical Model Across Distributed Storage

| Data Partitions | Data Partitions | Data Partitions | Data Partitions |

Files Across S3, Azure Blob, or HDFS

Figure 1-5. Data is distributed across physical machines

Partitioning allows for efficient parallelism. A distributed scheme of breaking up data into chunks or partitions allows Spark executors to process only data that is close to them, minimizing network bandwidth. That is, each executor's core is assigned its own data partition to work on (see Figure 1-6).

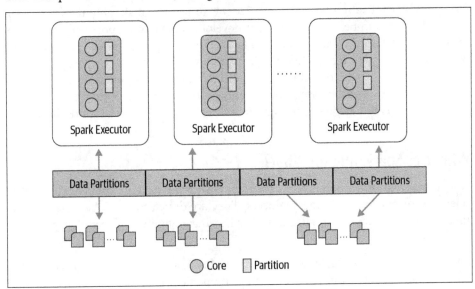

Figure 1-6. Each executor's core gets a partition of data to work on

For example, this code snippet will break up the physical data stored across clusters into eight partitions, and each executor will get one or more partitions to read into its memory:

```
# In Python
log_df = spark.read.text("path_to_large_text_file").repartition(8)
print(log_df.rdd.getNumPartitions())
```

And this code will create a DataFrame of 10,000 integers distributed over eight partitions in memory:

```
# In Python
df = spark.range(0, 10000, 1, 8)
print(df.rdd.getNumPartitions())
```

Both code snippets will print out 8.

In Chapters 3 and 7, we will discuss how to tune and change partitioning configuration for maximum parallelism based on how many cores you have on your executors.

The Developer's Experience

Of all the developers' delights, none is more attractive than a set of composable APIs that increase productivity and are easy to use, intuitive, and expressive. One of Apache Spark's principal appeals to developers has been its easy-to-use APIs (*https://oreil.ly/80dKh*) for operating on small to large data sets, across languages: Scala, Java, Python, SQL, and R.

One primary motivation behind Spark 2.x was to unify and simplify the framework by limiting the number of concepts that developers have to grapple with. Spark 2.x introduced higher-level abstraction APIs as domain-specific language constructs, which made programming Spark highly expressive and a pleasant developer experience. You express what you want the task or operation to compute, not how to compute it, and let Spark ascertain how best to do it for you. We will cover these Structured APIs in Chapter 3, but first let's take a look at who the Spark developers are.

Who Uses Spark, and for What?

Not surprisingly, most developers who grapple with big data are data engineers, data scientists, or machine learning engineers. They are drawn to Spark because it allows them to build a range of applications using a single engine, with familiar programming languages.

Of course, developers may wear many hats and sometimes do both data science and data engineering tasks, especially in startup companies or smaller engineering groups. Among all these tasks, however, data—massive amounts of data—is the foundation.

Data science tasks

As a discipline that has come to prominence in the era of big data, data science is about using data to tell stories. But before they can narrate the stories, data scientists have to cleanse the data, explore it to discover patterns, and build models to predict or suggest outcomes. Some of these tasks require knowledge of statistics, mathematics, computer science, and programming.

Most data scientists are proficient in using analytical tools like SQL, comfortable with libraries like NumPy and pandas, and conversant in programming languages like R and Python. But they must also know how to *wrangle* or *transform* data, and how to use established classification, regression, or clustering algorithms for building models. Often their tasks are iterative, interactive or ad hoc, or experimental to assert their hypotheses.

Fortunately, Spark supports these different tools. Spark's MLlib offers a common set of machine learning algorithms to build model pipelines, using high-level estimators, transformers, and data featurizers. Spark SQL and the Spark shell facilitate interactive and ad hoc exploration of data.

Additionally, Spark enables data scientists to tackle large data sets and scale their model training and evaluation. Apache Spark 2.4 introduced a new gang scheduler, as part of Project Hydrogen (*https://oreil.ly/8h3wr*), to accommodate the fault-tolerant needs of training and scheduling deep learning models in a distributed manner, and Spark 3.0 has introduced the ability to support GPU resource collection in the standalone, YARN, and Kubernetes deployment modes. This means developers whose tasks demand deep learning techniques can use Spark.

Data engineering tasks

After building their models, data scientists often need to work with other team members, who may be responsible for deploying the models. Or they may need to work closely with others to build and transform raw, dirty data into clean data that is easily consumable or usable by other data scientists. For example, a classification or clustering model does not exist in isolation; it works in conjunction with other components like a web application or a streaming engine such as Apache Kafka, or as part of a larger data pipeline. This pipeline is often built by data engineers.

Data engineers have a strong understanding of software engineering principles and methodologies, and possess skills for building scalable data pipelines for a stated business use case. Data pipelines enable end-to-end transformations of raw data coming from myriad sources—data is cleansed so that it can be consumed downstream by developers, stored in the cloud or in NoSQL or RDBMSs for report generation, or made accessible to data analysts via business intelligence tools.

Spark 2.x introduced an evolutionary streaming model called *continuous applications* (*https://oreil.ly/p0_fC*) with Structured Streaming (discussed in detail in Chapter 8). With Structured Streaming APIs, data engineers can build complex data pipelines that enable them to ETL data from both real-time and static data sources.

Data engineers use Spark because it provides a simple way to parallelize computations and hides all the complexity of distribution and fault tolerance. This leaves them free to focus on using high-level DataFrame-based APIs and domain-specific language (DSL) queries to do ETL, reading and combining data from multiple sources.

The performance improvements in Spark 2.x and Spark 3.0, due to the Catalyst optimizer (*https://oreil.ly/pAHKJ*) for SQL and Tungsten (*https://oreil.ly/nIE6h*) for compact code generation, have made life for data engineers much easier. They can choose to use any of the three Spark APIs (*https://oreil.ly/c1sf8*)—RDDs, DataFrames, or Datasets—that suit the task at hand, and reap the benefits of Spark.

Popular Spark use cases

Whether you are a data engineer, data scientist, or machine learning engineer, you'll find Spark useful for the following use cases:

- Processing in parallel large data sets distributed across a cluster
- Performing ad hoc or interactive queries to explore and visualize data sets
- Building, training, and evaluating machine learning models using MLlib
- Implementing end-to-end data pipelines from myriad streams of data
- Analyzing graph data sets and social networks

Community Adoption and Expansion

Not surprisingly, Apache Spark struck a chord in the open source community, especially among data engineers and data scientists. Its design philosophy and its inclusion as an Apache Software Foundation project have fostered immense interest among the developer community.

Today, there are over 600 Apache Spark Meetup groups (*https://oreil.ly/XjqQN*) globally with close to half a million members. Every week, someone in the world is giving a talk at a meetup or conference or sharing a blog post on how to use Spark to build data pipelines. The Spark + AI Summit (*https://oreil.ly/G9vYT*) is the largest conference dedicated to the use of Spark for machine learning, data engineering, and data science across many verticals.

Since Spark's first 1.0 release in 2014 there have been many minor and major releases, with the most recent major release of Spark 3.0 coming in 2020. This book will cover aspects of Spark 2.x and Spark 3.0. By the time of its publication the community will have released Spark 3.0, and most of the code in this book has been tested with Spark 3.0-preview2.

Over the course of its releases, Spark has continued to attract contributors from across the globe and from numerous organizations. Today, Spark has close to 1,500 contributors, well over 100 releases, 21,000 forks, and some 27,000 commits on GitHub, as Figure 1-7 shows. And we hope that when you finish this book, you will feel compelled to contribute too.

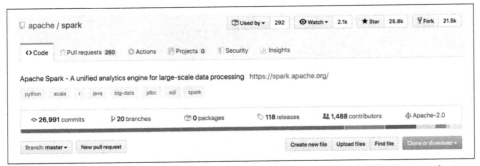

Figure 1-7. The state of Apache Spark on GitHub (source: https://github.com/apache/ spark)

Now we can turn our attention to the fun of learning—where and how to start using Spark. In the next chapter, we'll show you how to get up and running with Spark in three simple steps.

Downloading Apache Spark and Getting Started

In this chapter, we will get you set up with Spark and walk through three simple steps you can take to get started writing your first standalone application.

We will use local mode, where all the processing is done on a single machine in a Spark shell—this is an easy way to learn the framework, providing a quick feedback loop for iteratively performing Spark operations. Using a Spark shell, you can proto-type Spark operations with small data sets before writing a complex Spark application, but for large data sets or real work where you want to reap the benefits of distributed execution, local mode is not suitable—you'll want to use the YARN or Kubernetes deployment modes instead.

While the Spark shell only supports Scala, Python, and R, you can write a Spark application in any of the supported languages (including Java) and issue queries in Spark SQL. We do expect you to have some familiarity with the language of your choice.

Step 1: Downloading Apache Spark

To get started, go to the Spark download page (*https://oreil.ly/tbKY2*), select "Pre-built for Apache Hadoop 2.7" from the drop-down menu in step 2, and click the "Download Spark" link in step 3 (Figure 2-1).

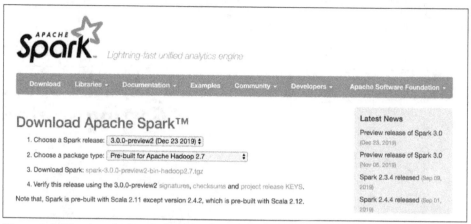

Figure 2-1. The Apache Spark download page

This will download the tarball *spark-3.0.0-preview2-bin-hadoop2.7.tgz*, which contains all the Hadoop-related binaries you will need to run Spark in local mode on your laptop. Alternatively, if you're going to install it on an existing HDFS or Hadoop installation, you can select the matching Hadoop version from the drop-down menu. How to build from source is beyond the scope of this book, but you can read more about it in the documentation (*https://oreil.ly/fOyIN*).

At the time this book went to press Apache Spark 3.0 was still in preview mode, but you can download the latest Spark 3.0 using the same download method and instructions.

Since the release of Apache Spark 2.2, developers who only care about learning Spark in Python have the option of installing PySpark from the PyPI repository (*https://oreil.ly/gyAi8*). If you only program in Python, you don't have to install all the other libraries necessary to run Scala, Java, or R; this makes the binary smaller. To install PySpark from PyPI, just run `pip install pyspark`.

There are some extra dependencies that can be installed for SQL, ML, and MLlib, via `pip install pyspark[sql,ml,mllib]` (or `pip install pyspark[sql]` if you only want the SQL dependencies).

You will need to install Java 8 or above on your machine and set the JAVA_HOME environment variable. See the documentation (*https://oreil.ly/c19W9*) for instructions on how to download and install Java.

If you want to run R in an interpretive shell mode, you must install R (*https://www.r-project.org*) and then run `sparkR`. To do distributed computing with R, you can also use the open source project `sparklyr` (*https://github.com/sparklyr/sparklyr*), created by the R community.

Spark's Directories and Files

We assume that you are running a version of the Linux or macOS operating system on your laptop or cluster, and all the commands and instructions in this book will be in that flavor. Once you have finished downloading the tarball, `cd` to the downloaded directory, extract the tarball contents with `tar -xf spark-3.0.0-preview2-bin-hadoop2.7.tgz`, and `cd` into that directory and take a look at the contents:

```
$ cd spark-3.0.0-preview2-bin-hadoop2.7
$ ls
LICENSE   R          RELEASE   conf   examples   kubernetes   python   yarn
NOTICE    README.md  bin       data   jars       licenses     sbin
```

Let's briefly summarize the intent and purpose of some of these files and directories. New items were added in Spark 2.x and 3.0, and the contents of some of the existing files and directories were changed too:

README.md

This file contains new detailed instructions on how to use Spark shells, build Spark from source, run standalone Spark examples, peruse links to Spark documentation and configuration guides, and contribute to Spark.

bin

This directory, as the name suggests, contains most of the scripts you'll employ to interact with Spark, including the Spark shells (`spark-sql`, `pyspark`, `spark-shell`, and `sparkR`). We will use these shells and executables in this directory later in this chapter to submit a standalone Spark application using `spark-submit`, and write a script that builds and pushes Docker images when running Spark with Kubernetes support.

sbin

Most of the scripts in this directory are administrative in purpose, for starting and stopping Spark components in the cluster in its various deployment modes. For details on the deployment modes, see the cheat sheet in Table 1-1 in Chapter 1.

kubernetes

Since the release of Spark 2.4, this directory contains Dockerfiles for creating Docker images for your Spark distribution on a Kubernetes cluster. It also contains a file providing instructions on how to build the Spark distribution before building your Docker images.

data

This directory is populated with **.txt* files that serve as input for Spark's components: MLlib, Structured Streaming, and GraphX.

examples

For any developer, two imperatives that ease the journey to learning any new platform are loads of "how-to" code examples and comprehensive documentation. Spark provides examples for Java, Python, R, and Scala, and you'll want to employ them when learning the framework. We will allude to some of these examples in this and subsequent chapters.

Step 2: Using the Scala or PySpark Shell

As mentioned earlier, Spark comes with four widely used interpreters that act like interactive "shells" and enable ad hoc data analysis: pyspark, spark-shell, spark-sql, and sparkR. In many ways, their interactivity imitates shells you'll already be familiar with if you have experience with Python, Scala, R, SQL, or Unix operating system shells such as bash or the Bourne shell.

These shells have been augmented to support connecting to the cluster and to allow you to load distributed data into Spark workers' memory. Whether you are dealing with gigabytes of data or small data sets, Spark shells are conducive to learning Spark quickly.

To start PySpark, cd to the *bin* directory and launch a shell by typing **pyspark**. If you have installed PySpark from PyPI, then just typing **pyspark** will suffice:

```
$ pyspark
Python 3.7.3 (default, Mar 27 2019, 09:23:15)
[Clang 10.0.1 (clang-1001.0.46.3)] on darwin
Type "help", "copyright", "credits" or "license" for more information.
20/02/16 19:28:48 WARN NativeCodeLoader: Unable to load native-hadoop library
for your platform... using builtin-java classes where applicable
Welcome to
      ____              __
     / __/__  ___ _____/ /__
    _\ \/ _ \/ _ `/ __/  '_/
   /__ / .__/\_,_/_/ /_/\_\   version 3.0.0-preview2
      /_/

Using Python version 3.7.3 (default, Mar 27 2019 09:23:15)
SparkSession available as 'spark'.
>>> spark.version
'3.0.0-preview2'
>>>
```

To start a similar Spark shell with Scala, cd to the *bin* directory and type
spark-shell:

```
$ spark-shell
20/05/07 19:30:26 WARN NativeCodeLoader: Unable to load native-hadoop library
for your platform... using builtin-java classes where applicable
Spark context Web UI available at http://10.0.1.7:4040
Spark context available as 'sc' (master = local[*], app id = local-1581910231902)
Spark session available as 'spark'.
Welcome to


     ____              __
    / __/__  ___ _____/ /__
   _\ \/ _ \/ _ `/ __/  '_/
  /___/ .__/\_,_/_/ /_/\_\   version 3.0.0-preview2
     /_/

Using Scala version 2.12.10 (Java HotSpot(TM) 64-Bit Server VM, Java 1.8.0_241)
Type in expressions to have them evaluated.
Type :help for more information.
scala> spark.version
res0: String = 3.0.0-preview2
scala>
```

Using the Local Machine

Now that you've downloaded and installed Spark on your local machine, for the
remainder of this chapter you'll be using Spark interpretive shells locally. That is,
Spark will be running in local mode.

Refer to Table 1-1 in Chapter 1 for a reminder of which components run where in local mode.

As noted in the previous chapter, Spark computations are expressed as operations.
These operations are then converted into low-level RDD-based bytecode as tasks,
which are distributed to Spark's executors for execution.

Let's look at a short example where we read in a text file as a DataFrame, show a sample of the strings read, and count the total number of lines in the file. This simple
example illustrates the use of the high-level Structured APIs, which we will cover in
the next chapter. The show(10, false) operation on the DataFrame only displays the
first 10 lines without truncating; by default the truncate Boolean flag is true. Here's
what this looks like in the Scala shell:

```
scala> val strings = spark.read.text("../README.md")
strings: org.apache.spark.sql.DataFrame = [value: string]

scala> strings.show(10, false)
+----------------------------------------------------------------------+
|value                                                                 |
+----------------------------------------------------------------------+
|# Apache Spark                                                        |
|                                                                      |
|Spark is a unified analytics engine for large-scale data processing. It|
|provides high-level APIs in Scala, Java, Python, and R, and an optimized|
|engine that supports general computation graphs for data analysis. It also|
|supports a rich set of higher-level tools including Spark SQL for SQL and|
|DataFrames, MLlib for machine learning, GraphX for graph processing,  |
| and Structured Streaming for stream processing.                      |
|                                                                      |
|<https://spark.apache.org/>                                           |
+----------------------------------------------------------------------+
only showing top 10 rows

scala> strings.count()
res2: Long = 109
scala>
```

Quite simple. Let's look at a similar example using the Python interpretive shell, pyspark:

```
$ pyspark
Python 3.7.3 (default, Mar 27 2019, 09:23:15)
[Clang 10.0.1 (clang-1001.0.46.3)] on darwin
Type "help", "copyright", "credits" or "license" for more information.
WARNING: An illegal reflective access operation has occurred
WARNING: Illegal reflective access by org.apache.spark.unsafe.Platform
WARNING: Use --illegal-access=warn to enable warnings of further illegal
reflective access operations
WARNING: All illegal access operations will be denied in a future release
20/01/10 11:28:29 WARN NativeCodeLoader: Unable to load native-hadoop library
for your platform... using builtin-java classes where applicable
Using Spark's default log4j profile: org/apache/spark/log4j-defaults.properties
Setting default log level to "WARN".
To adjust logging level use sc.setLogLevel(newLevel). For SparkR, use
setLogLevel(newLevel).
Welcome to
      ____              __
     / __/__  ___ _____/ /__
    _\ \/ _ \/ _ `/ __/  '_/
   /__ / .__/\_,_/_/ /_/\_\   version 3.0.0-preview2
      /_/

Using Python version 3.7.3 (default, Mar 27 2019 09:23:15)
SparkSession available as 'spark'.
>>> strings = spark.read.text("../README.md")
```

```
>>> strings.show(10, truncate=False)
+------------------------------------------------------------------------------+
|value                                                                         |
+------------------------------------------------------------------------------+
|# Apache Spark                                                                |
|                                                                              |
|Spark is a unified analytics engine for large-scale data processing. It       |
|provides high-level APIs in Scala, Java, Python, and R, and an optimized      |
|engine that supports general computation graphs for data analysis. It also    |
|supports a rich set of higher-level tools including Spark SQL for SQL and      |
|DataFrames, MLlib for machine learning, GraphX for graph processing,          |
|and Structured Streaming for stream processing.                               |
|                                                                              |
|<https://spark.apache.org/>                                                   |
+------------------------------------------------------------------------------+
only showing top 10 rows

>>> strings.count()
109
>>>
```

To exit any of the Spark shells, press Ctrl-D. As you can see, this rapid interactivity with Spark shells is conducive not only to rapid learning but to rapid prototyping, too.

In the preceding examples, notice the API syntax and signature parity across both Scala and Python. Throughout Spark's evolution from 1.x, that has been one (among many) of the enduring improvements.

Also note that we used the high-level Structured APIs to read a text file into a Spark DataFrame rather than an RDD. Throughout the book, we will focus more on these Structured APIs; since Spark 2.x, RDDs are now consigned to low-level APIs.

 Every computation expressed in high-level Structured APIs is decomposed into low-level optimized and generated RDD operations and then converted into Scala bytecode for the executors' JVMs. This generated RDD operation code is not accessible to users, nor is it the same as the user-facing RDD APIs.

Step 3: Understanding Spark Application Concepts

Now that you have downloaded Spark, installed it on your laptop in standalone mode, launched a Spark shell, and executed some short code examples interactively, you're ready to take the final step.

To understand what's happening under the hood with our sample code, you'll need to be familiar with some of the key concepts of a Spark application and how the code is transformed and executed as tasks across the Spark executors. We'll begin by defining some important terms:

Application

    A user program built on Spark using its APIs. It consists of a driver program and executors on the cluster.

SparkSession

    An object that provides a point of entry to interact with underlying Spark functionality and allows programming Spark with its APIs. In an interactive Spark shell, the Spark driver instantiates a SparkSession for you, while in a Spark application, you create a SparkSession object yourself.

Job

    A parallel computation consisting of multiple tasks that gets spawned in response to a Spark action (e.g., save(), collect()).

Stage

    Each job gets divided into smaller sets of tasks called stages that depend on each other.

Task

    A single unit of work or execution that will be sent to a Spark executor.

Let's dig into these concepts in a little more detail.

Spark Application and SparkSession

At the core of every Spark application is the Spark driver program, which creates a SparkSession object. When you're working with a Spark shell, the driver is part of the shell and the SparkSession object (accessible via the variable spark) is created for you, as you saw in the earlier examples when you launched the shells.

In those examples, because you launched the Spark shell locally on your laptop, all the operations ran locally, in a single JVM. But you can just as easily launch a Spark shell to analyze data in parallel on a cluster as in local mode. The commands spark-shell --help or pyspark --help will show you how to connect to the Spark cluster manager. Figure 2-2 shows how Spark executes on a cluster once you've done this.

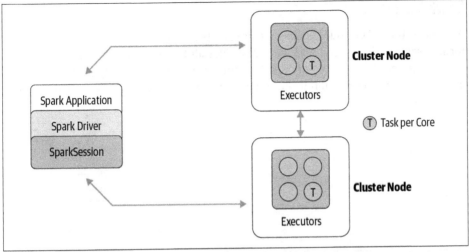

Figure 2-2. Spark components communicate through the Spark driver in Spark's distributed architecture

Once you have a `SparkSession`, you can program Spark using the APIs (*https:// oreil.ly/2r5Xo*) to perform Spark operations.

Spark Jobs

During interactive sessions with Spark shells, the driver converts your Spark application into one or more Spark jobs (Figure 2-3). It then transforms each job into a DAG. This, in essence, is Spark's execution plan, where each node within a DAG could be a single or multiple Spark stages.

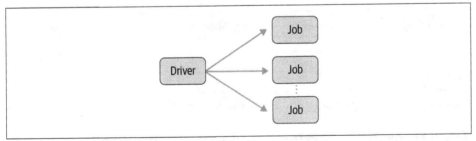

Figure 2-3. Spark driver creating one or more Spark jobs

Spark Stages

As part of the DAG nodes, stages are created based on what operations can be performed serially or in parallel (Figure 2-4). Not all Spark operations can happen in a single stage, so they may be divided into multiple stages. Often stages are delineated on the operator's computation boundaries, where they dictate data transfer among Spark executors.

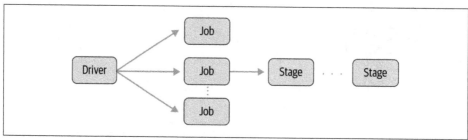

Figure 2-4. Spark job creating one or more stages

Spark Tasks

Each stage is comprised of Spark tasks (a unit of execution), which are then federated across each Spark executor; each task maps to a single core and works on a single partition of data (Figure 2-5). As such, an executor with 16 cores can have 16 or more tasks working on 16 or more partitions in parallel, making the execution of Spark's tasks exceedingly parallel!

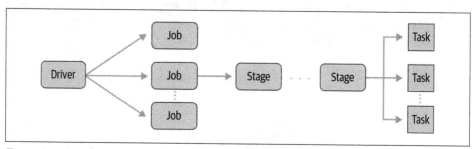

Figure 2-5. Spark stage creating one or more tasks to be distributed to executors

Transformations, Actions, and Lazy Evaluation

Spark operations on distributed data can be classified into two types: *transformations* and *actions*. Transformations, as the name suggests, transform a Spark DataFrame into a new DataFrame without altering the original data, giving it the property of immutability. Put another way, an operation such as `select()` or `filter()` will not change the original DataFrame; instead, it will return the transformed results of the operation as a new DataFrame.

All transformations are evaluated lazily. That is, their results are not computed imme-diately, but they are recorded or remembered as a *lineage*. A recorded lineage allows Spark, at a later time in its execution plan, to rearrange certain transformations, coa-lesce them, or optimize transformations into stages for more efficient execution. Lazy evaluation is Spark's strategy for delaying execution until an action is invoked or data is "touched" (read from or written to disk).

An action triggers the lazy evaluation of all the recorded transformations. In Figure 2-6, all transformations T are recorded until the action A is invoked. Each transformation T produces a new DataFrame.

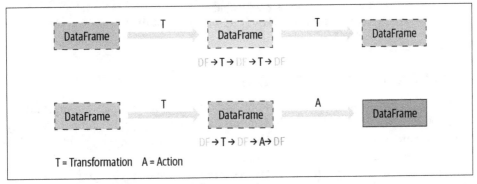

Figure 2-6. Lazy transformations and eager actions

While lazy evaluation allows Spark to optimize your queries by peeking into your chained transformations, lineage and data immutability provide fault tolerance. Because Spark records each transformation in its lineage and the DataFrames are immutable between transformations, it can reproduce its original state by simply replaying the recorded lineage, giving it resiliency in the event of failures.

Table 2-1 lists some examples of transformations and actions.

Table 2-1. Transformations and actions as Spark operations

Transformations	Actions
orderBy()	show()
groupBy()	take()
filter()	count()
select()	collect()
join()	save()

The actions and transformations contribute to a Spark query plan, which we will cover in the next chapter. Nothing in a query plan is executed until an action is invoked. The following example, shown both in Python and Scala, has two transfor-mations—read() and filter()—and one action—count(). The action is what triggers the execution of all transformations recorded as part of the query execution plan. In this example, nothing happens until `filtered.count()` is executed in the shell:

```python
# In Python
>>> strings = spark.read.text("../README.md")
>>> filtered = strings.filter(strings.value.contains("Spark"))
>>> filtered.count()
20
```

```scala
// In Scala
scala> import org.apache.spark.sql.functions._
scala> val strings = spark.read.text("../README.md")
scala> val filtered = strings.filter(col("value").contains("Spark"))
scala> filtered.count()
res5: Long = 20
```

Narrow and Wide Transformations

As noted, transformations are operations that Spark evaluates lazily. A huge advantage of the lazy evaluation scheme is that Spark can inspect your computational query and ascertain how it can optimize it. This optimization can be done by either joining or pipelining some operations and assigning them to a stage, or breaking them into stages by determining which operations require a shuffle or exchange of data across clusters.

Transformations can be classified as having either *narrow dependencies* or *wide dependencies*. Any transformation where a single output partition can be computed from a single input partition is a *narrow* transformation. For example, in the previous code snippet, `filter()` and `contains()` represent narrow transformations because they can operate on a single partition and produce the resulting output partition without any exchange of data.

However, transformations such as `groupBy()` or `orderBy()` instruct Spark to perform wide transformations, where data from other partitions is read in, combined, and written to disk. If we were to sort the `filtered` DataFrame from the preceding example by calling `.orderBy()`, each partition will be locally sorted, but we need to force a shuffle of data from each of the executor's partitions across the cluster to sort all of the records. In contrast to narrow transformations, wide transformations require output from other partitions to compute the final aggregation.

Figure 2-7 illustrates the two types of dependencies.

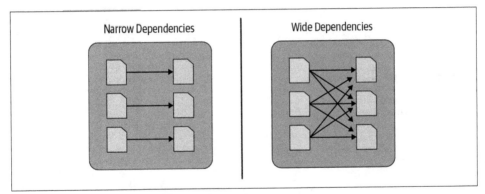

Figure 2-7. Narrow versus wide transformations

The Spark UI

Spark includes a graphical user interface (*https://oreil.ly/AXg5h*) that you can use to inspect or monitor Spark applications in their various stages of decomposition—that is jobs, stages, and tasks. Depending on how Spark is deployed, the driver launches a web UI, running by default on port 4040, where you can view metrics and details such as:

- A list of scheduler stages and tasks
- A summary of RDD sizes and memory usage
- Information about the environment
- Information about the running executors
- All the Spark SQL queries

In local mode, you can access this interface at *http://<localhost>:4040* in a web browser.

> When you launch `spark-shell`, part of the output shows the local-host URL to access at port 4040.

Let's inspect how the Python example from the previous section translates into jobs, stages, and tasks. To view what the DAG looks like, click on "DAG Visualization" in the web UI. As Figure 2-8 shows, the driver created a single job and a single stage.

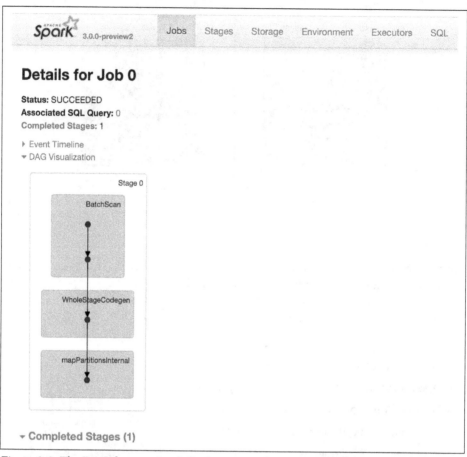

Figure 2-8. The DAG for our simple Python example

Notice that there is no Exchange, where data is exchanged between executors, required because there is only a single stage. The individual operations of the stage are shown in blue boxes.

Stage 0 is comprised of one task. If you have multiple tasks, they will be executed in parallel. You can view the details of each stage in the Stages tab, as shown in Figure 2-9.

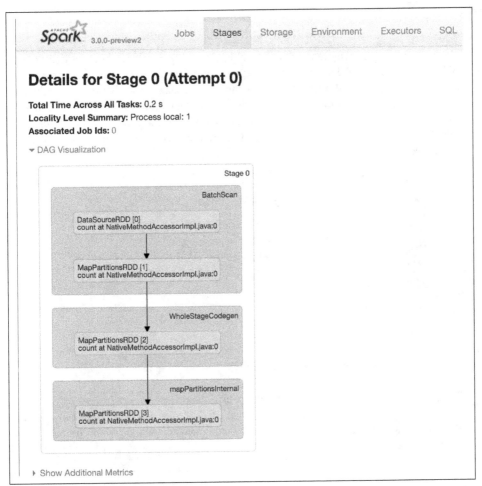

Figure 2-9. Details of stage 0

We will cover the Spark UI in more detail in Chapter 7. For now, just note that the UI provides a microscopic lens into Spark's internal workings as a tool for debugging and inspecting.

Databricks Community Edition

Databricks is a company that offers a managed Apache Spark platform in the cloud. Aside from using your local machine to run Spark in local mode, you can try some of the examples in this and other chapters using the free Databricks Community Edition (Figure 2-10). As a learning tool for Apache Spark, the Community Edition has many tutorials and examples worthy of note. As well as writing your own notebooks in Python, R, Scala, or SQL, you can also import other notebooks, including Jupyter notebooks.

Figure 2-10. Databricks Community Edition

To get an account, go to *https://www.databricks.com/try-databricks* and follow the instructions to try the Community Edition for free. Once registered, you can import the notebooks for this book from its GitHub repo (*https://github.com/databricks/LearningSparkV2*).

Your First Standalone Application

To facilitate learning and exploring, the Spark distribution comes with a set of sample applications for each of Spark's components. You are welcome to peruse the *examples* directory in your installation location to get an idea of what's available.

From the installation directory on your local machine, you can run one of the several Java or Scala sample programs that are provided using the command `bin/run-example <class> [params]`. For example:

```
$ ./bin/run-example JavaWordCount README.md
```

This will spew out INFO messages on your console along with a list of each word in the *README.md* file and its count (counting words is the "Hello, World" of distributed computing).

Counting M&Ms for the Cookie Monster

In the previous example, we counted words in a file. If the file were huge, it would be distributed across a cluster partitioned into small chunks of data, and our Spark program would distribute the task of counting each word in each partition and return us the final aggregated count. But that example has become a bit of a cliche.

Let's solve a similar problem, but with a larger data set and using more of Spark's distribution functionality and DataFrame APIs. We will cover the APIs used in this program in later chapters, but for now bear with us.

Among the authors of this book is a data scientist who loves to bake cookies with M&Ms in them, and she rewards her students in the US states where she frequently teaches machine learning and data science courses with batches of those cookies. But she's data-driven, obviously, and wants to ensure that she gets the right colors of M&Ms in the cookies for students in the different states (Figure 2-11).

Figure 2-11. Distribution of M&Ms by color (source: https://oreil.ly/mhWIT)

Let's write a Spark program that reads a file with over 100,000 entries (where each row or line has a `<state, mnm_color, count>`) and computes and aggregates the counts for each color and state. These aggregated counts tell us the colors of M&Ms favored by students in each state. The complete Python listing is provided in Example 2-1.

Example 2-1. Counting and aggregating M&Ms (Python version)

```python
# Import the necessary libraries.
# Since we are using Python, import the SparkSession and related functions
# from the PySpark module.
import sys

from pyspark.sql import SparkSession

if __name__ == "__main__":
    if len(sys.argv) != 2:
        print("Usage: mnmcount <file>", file=sys.stderr)
        sys.exit(-1)

    # Build a SparkSession using the SparkSession APIs.
    # If one does not exist, then create an instance. There
    # can only be one SparkSession per JVM.
    spark = (SparkSession
      .builder
      .appName("PythonMnMCount")
      .getOrCreate())
    # Get the M&M data set filename from the command-line arguments
    mnm_file = sys.argv[1]
    # Read the file into a Spark DataFrame using the CSV
    # format by inferring the schema and specifying that the
    # file contains a header, which provides column names for comma-
    # separated fields.
    mnm_df = (spark.read.format("csv")
      .option("header", "true")
      .option("inferSchema", "true")
      .load(mnm_file))

    # We use the DataFrame high-level APIs. Note
    # that we don't use RDDs at all. Because some of Spark's
    # functions return the same object, we can chain function calls.
    # 1. Select from the DataFrame the fields "State", "Color", and "Count"
    # 2. Since we want to group each state and its M&M color count,
    #    we use groupBy()
    # 3. Aggregate counts of all colors and groupBy() State and Color
    # 4  orderBy() in descending order
    count_mnm_df = (mnm_df
      .select("State", "Color", "Count")
      .groupBy("State", "Color")
      .sum("Count")
      .orderBy("sum(Count)", ascending=False))
    # Show the resulting aggregations for all the states and colors;
    # a total count of each color per state.
    # Note show() is an action, which will trigger the above
    # query to be executed.
    count_mnm_df.show(n=60, truncate=False)
    print("Total Rows = %d" % (count_mnm_df.count()))
    # While the above code aggregated and counted for all
```

```
# the states, what if we just want to see the data for
# a single state, e.g., CA?
# 1. Select from all rows in the DataFrame
# 2. Filter only CA state
# 3. groupBy() State and Color as we did above
# 4. Aggregate the counts for each color
# 5. orderBy() in descending order
# Find the aggregate count for California by filtering
ca_count_mnm_df = (mnm_df
  .select("State", "Color", "Count")
  .where(mnm_df.State == "CA")
  .groupBy("State", "Color")
  .sum("Count")
  .orderBy("sum(Count)", ascending=False))
# Show the resulting aggregation for California.
# As above, show() is an action that will trigger the execution of the
# entire computation.
ca_count_mnm_df.show(n=10, truncate=False)
# Stop the SparkSession
spark.stop()
```

You can enter this code into a Python file called *mnmcount.py* using your favorite editor, download the *mnn_dataset.csv* file from this book's GitHub repo (*https://github.com/databricks/LearningSparkV2*), and submit it as a Spark job using the submit-spark script in the installation's *bin* directory. Set your SPARK_HOME environment variable to the root-level directory where you installed Spark on your local machine.

 The preceding code uses the DataFrame API, which reads like high-level DSL queries. We will cover this and the other APIs in the next chapter; for now, note the clarity and simplicity with which you can instruct Spark what to do, not how to do it, unlike with the RDD API. Cool stuff!

To avoid having verbose INFO messages printed to the console, copy the *log4j.properties.template* file to *log4j.properties* and set log4j.rootCategory=WARN in the *conf/log4j.properties* file.

Let's submit our first Spark job using the Python APIs (for an explanation of what the code does, please read the inline comments in Example 2-1):

```
$SPARK_HOME/bin/spark-submit mnmcount.py data/mnm_dataset.csv
```

```
-----+------+----------+
|State| Color|sum(Count)|
+-----+------+----------+
|   CA|Yellow|    100956|
|   WA| Green|     96486|
|   CA| Brown|     95762|
```

```
|  TX|  Green|    95753|
|  TX|    Red|    95404|
|  CO|Yellow|    95038|
|  NM|    Red|    94699|
|  OR|Orange|    94514|
|  WY|  Green|    94339|
|  NV|Orange|    93929|
|  TX|Yellow|    93819|
|  CO|  Green|    93724|
|  CO|  Brown|    93692|
|  CA|  Green|    93505|
|  NM|  Brown|    93447|
|  CO|   Blue|    93412|
|  WA|    Red|    93332|
|  WA|  Brown|    93082|
|  WA|Yellow|    92920|
|  NM|Yellow|    92747|
|  NV|  Brown|    92478|
|  TX|Orange|    92315|
|  AZ|  Brown|    92287|
|  AZ|  Green|    91882|
|  WY|    Red|    91768|
|  AZ|Orange|    91684|
|  CA|    Red|    91527|
|  WA|Orange|    91521|
|  NV|Yellow|    91390|
|  UT|Orange|    91341|
|  NV|  Green|    91331|
|  NM|Orange|    91251|
|  NM|  Green|    91160|
|  WY|   Blue|    91002|
|  UT|    Red|    90995|
|  CO|Orange|    90971|
|  AZ|Yellow|    90946|
|  TX|  Brown|    90736|
|  OR|   Blue|    90526|
|  CA|Orange|    90311|
|  OR|    Red|    90286|
|  NM|   Blue|    90150|
|  AZ|    Red|    90042|
|  NV|   Blue|    90003|
|  UT|   Blue|    89977|
|  AZ|   Blue|    89971|
|  WA|   Blue|    89886|
|  OR|  Green|    89578|
|  CO|    Red|    89465|
|  NV|    Red|    89346|
|  UT|Yellow|    89264|
|  OR|  Brown|    89136|
|  CA|   Blue|    89123|
|  UT|  Brown|    88973|
|  TX|   Blue|    88466|
```

```
|  UT|  Green|     88392|
|  OR|Yellow|     88129|
|  WY|Orange|     87956|
|  WY|Yellow|     87800|
|  WY|  Brown|     86110|
+-----+------+----------+

Total Rows = 60

+-----+------+----------+
|State| Color|sum(Count)|
+-----+------+----------+
|   CA|Yellow|    100956|
|   CA| Brown|     95762|
|   CA| Green|     93505|
|   CA|   Red|     91527|
|   CA|Orange|     90311|
|   CA|  Blue|     89123|
+-----+------+----------+
```

First we see all the aggregations for each M&M color for each state, followed by those only for CA (where the preferred color is yellow).

What if you want to use a Scala version of this same Spark program? The APIs are similar; in Spark, parity is well preserved across the supported languages, with minor syntax differences. Example 2-2 is the Scala version of the program. Take a look, and in the next section we'll show you how to build and run the application.

Example 2-2. Counting and aggregating M&Ms (Scala version)

```scala
package main.scala.chapter2

import org.apache.spark.sql.SparkSession
import org.apache.spark.sql.functions._

/**
 * Usage: MnMcount <mnm_file_dataset>
 */
object MnMcount {
 def main(args: Array[String]) {
   val spark = SparkSession
     .builder
     .appName("MnMCount")
     .getOrCreate()

   if (args.length < 1) {
     print("Usage: MnMcount <mnm_file_dataset>")
     sys.exit(1)
   }
   // Get the M&M data set filename
   val mnmFile = args(0)
```

```scala
      // Read the file into a Spark DataFrame
      val mnmDF = spark.read.format("csv")
        .option("header", "true")
        .option("inferSchema", "true")
        .load(mnmFile)
      // Aggregate counts of all colors and groupBy() State and Color
      // orderBy() in descending order
      val countMnMDF = mnmDF
        .select("State", "Color", "Count")
        .groupBy("State", "Color")
        .sum("Count")
        .orderBy(desc("sum(Count)"))
      // Show the resulting aggregations for all the states and colors
      countMnMDF.show(60)
      println(s"Total Rows = ${countMnMDF.count()}")
      println()
      // Find the aggregate counts for California by filtering
      val caCountMnNDF = mnmDF
        .select("State", "Color", "Count")
        .where(col("State") === "CA")
        .groupBy("State", "Color")
        .sum("Count")
        .orderBy(desc("sum(Count)"))
      // Show the resulting aggregations for California
      caCountMnMDF.show(10)
      // Stop the SparkSession
      spark.stop()
  }
}
```

Building Standalone Applications in Scala

We will now show you how to build your first Scala Spark program, using the Scala Build Tool (sbt) (*https://www.scala-sbt.org*).

 Because Python is an interpreted language and there is no such step as compiling first (though it's possible to compile your Python code into bytecode in *.pyc*), we will not go into this step here. For details on how to use Maven to build Java Spark programs, we refer you to the guide (*https://oreil.ly/1qMIG*) on the Apache Spark website. For brevity in this book, we cover examples mainly in Python and Scala.

build.sbt is the specification file that, like a makefile, describes and instructs the Scala compiler to build your Scala-related tasks, such as jars, packages, what dependencies to resolve, and where to look for them. In our case, we have a simple sbt file for our M&M code (Example 2-3).

Example 2-3. sbt build file

```
// Name of the package
name := "main/scala/chapter2"
// Version of our package
version := "1.0"
// Version of Scala
scalaVersion := "2.12.10"
// Spark library dependencies
libraryDependencies ++= Seq(
  "org.apache.spark" %% "spark-core" % "3.0.0-preview2",
  "org.apache.spark" %% "spark-sql"  % "3.0.0-preview2"
)
```

Assuming that you have the Java Development Kit (JDK) (*https://oreil.ly/AfpMz*) and sbt installed and JAVA_HOME and SPARK_HOME set, with a single command, you can build your Spark application:

```
$ sbt clean package
[info] Updated file /Users/julesdamji/gits/LearningSparkV2/chapter2/scala/
project/build.properties: set sbt.version to 1.2.8
[info] Loading project definition from /Users/julesdamji/gits/LearningSparkV2/
chapter2/scala/project
[info] Updating
[info] Done updating.
...
[info] Compiling 1 Scala source to /Users/julesdamji/gits/LearningSparkV2/
chapter2/scala/target/scala-2.12/classes ...
[info] Done compiling.
[info] Packaging /Users/julesdamji/gits/LearningSparkV2/chapter2/scala/target/
scala-2.12/main-scala-chapter2_2.12-1.0.jar ...
[info] Done packaging.
[success] Total time: 6 s, completed Jan 11, 2020, 4:11:02 PM
```

After a successful build, you can run the Scala version of the M&M count example as follows:

```
$SPARK_HOME/bin/spark-submit --class main.scala.chapter2.MnMcount \
jars/main-scala-chapter2_2.12-1.0.jar data/mnm_dataset.csv
...
...
20/01/11 16:00:48 INFO TaskSchedulerImpl: Killing all running tasks in stage 4:
Stage finished
20/01/11 16:00:48 INFO DAGScheduler: Job 4 finished: show at MnMcount.scala:49,
took 0.264579 s
+-----+------+-----+
|State| Color|Total|
+-----+------+-----+
|   CA|Yellow| 1807|
|   CA| Green| 1723|
|   CA| Brown| 1718|
|   CA|Orange| 1657|
```

```
|  CA|   Red| 1656|
|  CA|  Blue| 1603|
+-----+------+-----+
```

The output is the same as for the Python run. Try it!

There you have it—our data scientist author will be more than happy to use this data to decide what colors of M&Ms to use in the cookies she bakes for her classes in any of the states she teaches in.

Summary

In this chapter, we covered the three simple steps you need to take to get started with Apache Spark: downloading the framework, familiarizing yourself with the Scala or PySpark interactive shell, and getting to grips with high-level Spark application concepts and terms. We gave a quick overview of the process by which you can use transformations and actions to write a Spark application, and we briefly introduced using the Spark UI to examine the jobs, stages, and tasks created.

Finally, through a short example, we showed you how you can use the high-level Structured APIs to tell Spark what to do—which brings us to the next chapter, where we examine those APIs in more detail.

## CHAPTER 3

Apache Spark's Structured APIs

In this chapter, we will explore the principal motivations behind adding structure to Apache Spark, how those motivations led to the creation of high-level APIs (Data-Frames and Datasets), and their unification in Spark 2.x across its components. We'll also look at the Spark SQL engine that underpins these structured high-level APIs.

When Spark SQL (*https://oreil.ly/cfd1r*) was first introduced in the early Spark 1.x releases, followed by DataFrames (*https://oreil.ly/kErKh*) as a successor to Sche-maRDDs (*https://oreil.ly/-o1-k*) in Spark 1.3, we got our first glimpse of structure in Spark. Spark SQL introduced high-level expressive operational functions, mimicking SQL-like syntax, and DataFrames, which laid the foundation for more structure in subsequent releases, paved the path to performant operations in Spark's computational queries.

But before we talk about the newer Structured APIs, let's get a brief glimpse of what it's like to not have structure in Spark by taking a peek at the simple RDD programming API model.

Spark: What's Underneath an RDD?

The RDD (*https://oreil.ly/KON5Y*) is the most basic abstraction in Spark. There are three vital characteristics associated with an RDD:

- Dependencies
- Partitions (with some locality information)
- Compute function: Partition => `Iterator[T]`

All three are integral to the simple RDD programming API model upon which all higher-level functionality is constructed. First, a list of *dependencies* that instructs Spark how an RDD is constructed with its inputs is required. When necessary to reproduce results, Spark can recreate an RDD from these dependencies and replicate operations on it. This characteristic gives RDDs resiliency.

Second, *partitions* provide Spark the ability to split the work to parallelize computation on partitions across executors. In some cases—for example, reading from HDFS—Spark will use locality information to send work to executors close to the data. That way less data is transmitted over the network.

And finally, an RDD has a *compute function* that produces an `Iterator[T]` for the data that will be stored in the RDD.

Simple and elegant! Yet there are a couple of problems with this original model. For one, the compute function (or computation) is opaque to Spark. That is, Spark does not know what you are doing in the compute function. Whether you are performing a join, filter, select, or aggregation, Spark only sees it as a lambda expression. Another problem is that the `Iterator[T]` data type is also opaque for Python RDDs; Spark only knows that it's a generic object in Python.

Furthermore, because it's unable to inspect the computation or expression in the function, Spark has no way to optimize the expression—it has no comprehension of its intention. And finally, Spark has no knowledge of the specific data type in T. To Spark it's an opaque object; it has no idea if you are accessing a column of a certain type within an object. Therefore, all Spark can do is serialize the opaque object as a series of bytes, without using any data compression techniques.

This opacity clearly hampers Spark's ability to rearrange your computation into an efficient query plan. So what's the solution?

Structuring Spark

Spark 2.x introduced a few key schemes for structuring Spark. One is to express computations by using common patterns found in data analysis. These patterns are expressed as high-level operations such as filtering, selecting, counting, aggregating, averaging, and grouping. This provides added clarity and simplicity.

This specificity is further narrowed through the use of a set of common operators in a DSL. Through a set of operations in DSL, available as APIs in Spark's supported languages (Java, Python, Spark, R, and SQL), these operators let you tell Spark what you wish to compute with your data, and as a result, it can construct an efficient query plan for execution.

And the final scheme of order and structure is to allow you to arrange your data in a tabular format, like a SQL table or spreadsheet, with supported structured data types (which we will cover shortly).

But what's all this structure good for?

Key Merits and Benefits

Structure yields a number of benefits, including better performance and space efficiency across Spark components. We will explore these benefits further when we talk about the use of the DataFrame and Dataset APIs shortly, but for now we'll concentrate on the other advantages: expressivity, simplicity, composability, and uniformity.

Let's demonstrate expressivity and composability first, with a simple code snippet. In the following example, we want to aggregate all the ages for each name, group by name, and then average the ages—a common pattern in data analysis and discovery. If we were to use the low-level RDD API for this, the code would look as follows:

```
# In Python
# Create an RDD of tuples (name, age)
dataRDD = sc.parallelize([("Brooke", 20), ("Denny", 31), ("Jules", 30),
  ("TD", 35), ("Brooke", 25)])
# Use map and reduceByKey transformations with their lambda
# expressions to aggregate and then compute average

agesRDD = (dataRDD
  .map(lambda x: (x[0], (x[1], 1)))
  .reduceByKey(lambda x, y: (x[0] + y[0], x[1] + y[1]))
  .map(lambda x: (x[0], x[1][0]/x[1][1])))
```

No one would dispute that this code, which tells Spark *how to* aggregate keys and compute averages with a string of lambda functions, is cryptic and hard to read. In other words, the code is instructing Spark how to compute the query. It's completely opaque to Spark, because it doesn't communicate the intention. Furthermore, the equivalent RDD code in Scala would look very different from the Python code shown here.

By contrast, what if we were to express the same query with high-level DSL operators and the DataFrame API, thereby instructing Spark *what to do*? Have a look:

```
# In Python
from pyspark.sql import SparkSession
from pyspark.sql.functions import avg
# Create a DataFrame using SparkSession
spark = (SparkSession
  .builder
  .appName("AuthorsAges")
  .getOrCreate())
# Create a DataFrame
data_df = spark.createDataFrame([("Brooke", 20), ("Denny", 31), ("Jules", 30),
```

```
  ("TD", 35), ("Brooke", 25)], ["name", "age"])
# Group the same names together, aggregate their ages, and compute an average
avg_df = data_df.groupBy("name").agg(avg("age"))
# Show the results of the final execution
avg_df.show()

+------+--------+
|  name|avg(age)|
+------+--------+
|Brooke|    22.5|
| Jules|    30.0|
|    TD|    35.0|
| Denny|    31.0|
+------+--------+
```

This version of the code is far more expressive as well as simpler than the earlier version, because we are using high-level DSL operators and APIs to tell Spark what to do. In effect, we have employed these operators to compose our query. And because Spark can inspect or parse this query and understand our intention, it can optimize or arrange the operations for efficient execution. Spark knows exactly *what* we wish to do: group people by their names, aggregate their ages, and then compute the average age of all people with the same name. We've composed an entire computation using high-level operators as a single simple query—how expressive is that?

Some would contend that by using only high-level, expressive DSL operators mapped to common or recurring data analysis patterns to introduce order and structure, we are limiting the scope of the developers' ability to instruct the compiler or control how their queries should be computed. Rest assured that you are not confined to these structured patterns; you can switch back at any time to the unstructured low-level RDD API, although we hardly ever find a need to do so.

As well as being simpler to read, the structure of Spark's high-level APIs also introduces uniformity across its components and languages. For example, the Scala code shown here does the same thing as the previous Python code—and the API looks nearly identical:

```
// In Scala
import org.apache.spark.sql.functions.avg
import org.apache.spark.sql.SparkSession
// Create a DataFrame using SparkSession
val spark = SparkSession
  .builder
  .appName("AuthorsAges")
  .getOrCreate()
// Create a DataFrame of names and ages
val dataDF = spark.createDataFrame(Seq(("Brooke", 20), ("Brooke", 25),
  ("Denny", 31), ("Jules", 30), ("TD", 35))).toDF("name", "age")
// Group the same names together, aggregate their ages, and compute an average
val avgDF = dataDF.groupBy("name").agg(avg("age"))
// Show the results of the final execution
```

```
avgDF.show()

+------+--------+
|  name|avg(age)|
+------+--------+
|Brooke|    22.5|
| Jules|    30.0|
|    TD|    35.0|
| Denny|    31.0|
+------+--------+
```

 Some of these DSL operators perform relational-like operations that you'll be familiar with if you know SQL, such as selecting, filtering, grouping, and aggregation.

All of this simplicity and expressivity that we developers cherish is possible because of the Spark SQL engine upon which the high-level Structured APIs are built. It is because of this engine, which underpins all the Spark components, that we get uniform APIs. Whether you express a query against a DataFrame in Structured Streaming or MLlib, you are always transforming and operating on DataFrames as structured data. We'll take a closer look at the Spark SQL engine later in this chapter, but for now let's explore those APIs and DSLs for common operations and how to use them for data analytics.

The DataFrame API

Inspired by pandas DataFrames (*https://oreil.ly/z93hD*) in structure, format, and a few specific operations, Spark DataFrames are like distributed in-memory tables with named columns and schemas, where each column has a specific data type: integer, string, array, map, real, date, timestamp, etc. To a human's eye, a Spark DataFrame is like a table. An example is shown in Table 3-1.

Table 3-1. The table-like format of a DataFrame

Id (Int)	First (String)	Last (String)	Url (String)	Published (Date)	Hits (Int)	Campaigns (List[Strings])
1	Jules	Damji	https://tinyurl.1	1/4/2016	4535	[twitter, LinkedIn]
2	Brooke	Wenig	https://tinyurl.2	5/5/2018	8908	[twitter, LinkedIn]
3	Denny	Lee	https://tinyurl.3	6/7/2019	7659	[web, twitter, FB, LinkedIn]
4	Tathagata	Das	https://tinyurl.4	5/12/2018	10568	[twitter, FB]

Id (Int)	First (String)	Last (String)	Url (String)	Published (Date)	Hits (Int)	Campaigns (List[Strings])
5	Matei	Zaharia	https:// tinyurl.5	5/14/2014	40578	[web, twitter, FB, LinkedIn]
6	Reynold	Xin	https:// tinyurl.6	3/2/2015	25568	[twitter, LinkedIn]

When data is visualized as a structured table, it's not only easy to digest but also easy to work with when it comes to common operations you might want to execute on rows and columns. Also recall that, as you learned in Chapter 2, DataFrames are immutable and Spark keeps a lineage of all transformations. You can add or change the names and data types of the columns, creating new DataFrames while the previous versions are preserved. A named column in a DataFrame and its associated Spark data type can be declared in the schema.

Let's examine the generic and structured data types available in Spark before we use them to define a schema. Then we'll illustrate how to create a DataFrame with a schema, capturing the data in Table 3-1.

Spark's Basic Data Types

Matching its supported programming languages, Spark supports basic internal data types. These data types can be declared in your Spark application or defined in your schema. For example, in Scala, you can define or declare a particular column name to be of type String, Byte, Long, or Map, etc. Here, we define variable names tied to a Spark data type:

```
$SPARK_HOME/bin/spark-shell
scala> import org.apache.spark.sql.types._
import org.apache.spark.sql.types._
scala> val nameTypes = StringType
nameTypes: org.apache.spark.sql.types.StringType.type = StringType
scala> val firstName = nameTypes
firstName: org.apache.spark.sql.types.StringType.type = StringType
scala> val lastName = nameTypes
lastName: org.apache.spark.sql.types.StringType.type = StringType
```

Table 3-2 lists the basic Scala data types supported in Spark. They all are subtypes of the class DataTypes (*https://oreil.ly/_GifO*), except for DecimalType.

Table 3-2. Basic Scala data types in Spark

Data type	Value assigned in Scala	API to instantiate
ByteType	Byte	DataTypes.ByteType
ShortType	Short	DataTypes.ShortType
IntegerType	Int	DataTypes.IntegerType
LongType	Long	DataTypes.LongType
FloatType	Float	DataTypes.FloatType
DoubleType	Double	DataTypes.DoubleType
StringType	String	DataTypes.StringType
BooleanType	Boolean	DataTypes.BooleanType
DecimalType	java.math.BigDecimal	DecimalType

Spark supports similar basic Python data types (*https://oreil.ly/HuREJ*), as enumerated in Table 3-3.

Table 3-3. Basic Python data types in Spark

Data type	Value assigned in Python	API to instantiate
ByteType	int	DataTypes.ByteType
ShortType	int	DataTypes.ShortType
IntegerType	int	DataTypes.IntegerType
LongType	int	DataTypes.LongType
FloatType	float	DataTypes.FloatType
DoubleType	float	DataTypes.DoubleType
StringType	str	DataTypes.StringType
BooleanType	bool	DataTypes.BooleanType
DecimalType	decimal.Decimal	DecimalType

Spark's Structured and Complex Data Types

For complex data analytics, you won't deal only with simple or basic data types. Your data will be complex, often structured or nested, and you'll need Spark to handle these complex data types. They come in many forms: maps, arrays, structs, dates, timestamps, fields, etc. Table 3-4 lists the Scala structured data types that Spark supports.

Table 3-4. Scala structured data types in Spark

Data type	Value assigned in Scala	API to instantiate
BinaryType	Array[Byte]	DataTypes.BinaryType
Timestamp Type	java.sql.Timestamp	DataTypes.TimestampType
DateType	java.sql.Date	DataTypes.DateType
ArrayType	scala.collection.Seq	DataTypes.createArrayType(Element Type)
MapType	scala.collection.Map	DataTypes.createMapType(keyType, valueType)
StructType	org.apache.spark.sql.Row	StructType(ArrayType[fieldTypes])
StructField	A value type corresponding to the type of this field	StructField(name, dataType, [nullable])

The equivalent structured data types in Python that Spark supports are enumerated in Table 3-5.

Table 3-5. Python structured data types in Spark

Data type	Value assigned in Python	API to instantiate
BinaryType	bytearray	BinaryType()
TimestampType	datetime.datetime	TimestampType()
DateType	datetime.date	DateType()
ArrayType	List, tuple, or array	ArrayType(dataType, [nullable])
MapType	dict	MapType(keyType, valueType, [nullable])
StructType	List or tuple	StructType([fields])
StructField	A value type corresponding to the type of this field	StructField(name, dataType, [nullable])

While these tables showcase the myriad types supported, it's far more important to see how these types come together when you define a schema for your data.

Schemas and Creating DataFrames

A *schema* in Spark defines the column names and associated data types for a Data-Frame. Most often, schemas come into play when you are reading structured data from an external data source (more on this in the next chapter). Defining a schema up front as opposed to taking a schema-on-read approach offers three benefits:

- You relieve Spark from the onus of inferring data types.
- You prevent Spark from creating a separate job just to read a large portion of your file to ascertain the schema, which for a large data file can be expensive and time-consuming.
- You can detect errors early if data doesn't match the schema.

So, we encourage you to always define your schema up front whenever you want to read a large file from a data source. For a short illustration, let's define a schema for the data in Table 3-1 and use that schema to create a DataFrame.

Two ways to define a schema

Spark allows you to define a schema in two ways. One is to define it programmatically, and the other is to employ a Data Definition Language (DDL) string, which is much simpler and easier to read.

To define a schema programmatically for a DataFrame with three named columns, author, title, and pages, you can use the Spark DataFrame API. For example:

```scala
// In Scala
import org.apache.spark.sql.types._
val schema = StructType(Array(StructField("author", StringType, false),
  StructField("title", StringType, false),
  StructField("pages", IntegerType, false)))
```

```python
# In Python
from pyspark.sql.types import *
schema = StructType([StructField("author", StringType(), False),
  StructField("title", StringType(), False),
  StructField("pages", IntegerType(), False)])
```

Defining the same schema using DDL is much simpler:

```scala
// In Scala
val schema = "author STRING, title STRING, pages INT"
```

```python
# In Python
schema = "author STRING, title STRING, pages INT"
```

You can choose whichever way you like to define a schema. For many examples, we will use both:

```python
# In Python
from pyspark.sql import SparkSession

# Define schema for our data using DDL
schema = "`Id` INT, `First` STRING, `Last` STRING, `Url` STRING,
  `Published` STRING, `Hits` INT, `Campaigns` ARRAY<STRING>"

# Create our static data
data = [[1, "Jules", "Damji", "https://tinyurl.1", "1/4/2016", 4535, ["twitter",
```

```
        "LinkedIn"]],
        [2, "Brooke","Wenig", "https://tinyurl.2", "5/5/2018", 8908, ["twitter",
"LinkedIn"]],
        [3, "Denny", "Lee",  "https://tinyurl.3", "6/7/2019", 7659, ["web",
"twitter", "FB", "LinkedIn"]],
        [4, "Tathagata", "Das", "https://tinyurl.4", "5/12/2018", 10568,
["twitter", "FB"]],
        [5, "Matei","Zaharia", "https://tinyurl.5", "5/14/2014", 40578, ["web",
"twitter", "FB", "LinkedIn"]],
        [6, "Reynold", "Xin", "https://tinyurl.6", "3/2/2015", 25568,
["twitter", "LinkedIn"]]
        ]

# Main program
if __name__ == "__main__":
    # Create a SparkSession
    spark = (SparkSession
        .builder
        .appName("Example-3_6")
        .getOrCreate())
    # Create a DataFrame using the schema defined above
    blogs_df = spark.createDataFrame(data, schema)
    # Show the DataFrame; it should reflect our table above
    blogs_df.show()
    # Print the schema used by Spark to process the DataFrame
    print(blogs_df.printSchema())
```

Running this program from the console will produce the following output:

```
$ spark-submit Example-3_6.py
...
+---+---------+-------+------------------+---------+-----+--------------------+
|Id |First    |Last   |Url               |Published|Hits |Campaigns           |
+---+---------+-------+------------------+---------+-----+--------------------+
|1  |Jules    |Damji  |https://tinyurl.1|1/4/2016 |4535 |[twitter,...]       |
|2  |Brooke   |Wenig  |https://tinyurl.2|5/5/2018 |8908 |[twitter,...]       |
|3  |Denny    |Lee    |https://tinyurl.3|6/7/2019 |7659 |[web, twitter...]   |
|4  |Tathagata|Das    |https://tinyurl.4|5/12/2018|10568|[twitter, FB]       |
|5  |Matei    |Zaharia|https://tinyurl.5|5/14/2014|40578|[web, twitter,...]| 
|6  |Reynold  |Xin    |https://tinyurl.6|3/2/2015 |25568|[twitter,...]       |
+---+---------+-------+------------------+---------+-----+--------------------+

root
 |-- Id: integer (nullable = false)
 |-- First: string (nullable = false)
 |-- Last: string (nullable = false)
 |-- Url: string (nullable = false)
 |-- Published: string (nullable = false)
 |-- Hits: integer (nullable = false)
 |-- Campaigns: array (nullable = false)
 |    |-- element: string (containsNull = false)
```

If you want to use this schema elsewhere in your code, simply execute `blogs_df.schema` and it will return the schema definition:

```
StructType(List(StructField("Id",IntegerType,false),
StructField("First",StringType,false),
StructField("Last",StringType,false),
StructField("Url",StringType,false),
StructField("Published",StringType,false),
StructField("Hits",IntegerType,false),
StructField("Campaigns",ArrayType(StringType,true),false)))
```

As you can observe, the DataFrame layout matches that of Table 3-1 along with the respective data types and schema output.

If you were to read the data from a JSON file instead of creating static data, the schema definition would be identical. Let's illustrate the same code with a Scala example, this time reading from a JSON file:

```scala
// In Scala
package main.scala.chapter3

import org.apache.spark.sql.SparkSession
import org.apache.spark.sql.types._

object Example3_7 {
 def main(args: Array[String]) {

   val spark = SparkSession
     .builder
     .appName("Example-3_7")
     .getOrCreate()

   if (args.length <= 0) {
     println("usage Example3_7 <file path to blogs.json>")
     System.exit(1)
   }
   // Get the path to the JSON file
   val jsonFile = args(0)
   // Define our schema programmatically
   val schema = StructType(Array(StructField("Id", IntegerType, false),
     StructField("First", StringType, false),
     StructField("Last", StringType, false),
     StructField("Url", StringType, false),
     StructField("Published", StringType, false),
     StructField("Hits", IntegerType, false),
     StructField("Campaigns", ArrayType(StringType), false)))

   // Create a DataFrame by reading from the JSON file
   // with a predefined schema
   val blogsDF = spark.read.schema(schema).json(jsonFile)
   // Show the DataFrame schema as output
   blogsDF.show(false)
```

```
    // Print the schema
    println(blogsDF.printSchema)
    println(blogsDF.schema)
  }
}
```

Not surprisingly, the output from the Scala program is no different than that from the Python program:

```
+---+---------+-------+-----------------+---------+-----+--------------------+
|Id |First    |Last   |Url              |Published|Hits |Campaigns           |
+---+---------+-------+-----------------+---------+-----+--------------------+
|1  |Jules    |Damji  |https://tinyurl.1|1/4/2016 |4535 |[twitter, LinkedIn] |
|2  |Brooke   |Wenig  |https://tinyurl.2|5/5/2018 |8908 |[twitter, LinkedIn] |
|3  |Denny    |Lee    |https://tinyurl.3|6/7/2019 |7659 |[web, twitter,...]  |
|4  |Tathagata|Das    |https://tinyurl.4|5/12/2018|10568|[twitter, FB]       |
|5  |Matei    |Zaharia|https://tinyurl.5|5/14/2014|40578|[web, twitter, FB,...]|
|6  |Reynold  |Xin    |https://tinyurl.6|3/2/2015 |25568|[twitter, LinkedIn] |
+---+---------+-------+-----------------+---------+-----+--------------------+

root
 |-- Id: integer (nullable = true)
 |-- First: string (nullable = true)
 |-- Last: string (nullable = true)
 |-- Url: string (nullable = true)
 |-- Published: string (nullable = true)
 |-- Hits: integer (nullable = true)
 |-- Campaigns: array (nullable = true)
 |    |-- element: string (containsNull = true)

StructType(StructField("Id",IntegerType,true),
    StructField("First",StringType,true),
    StructField("Last",StringType,true),
    StructField("Url",StringType,true),
    StructField("Published",StringType,true),
    StructField("Hits",IntegerType,true),
    StructField("Campaigns",ArrayType(StringType,true),true))
```

Now that you have an idea of how to use structured data and schemas in DataFrames, let's focus on DataFrame columns and rows and what it means to operate on them with the DataFrame API.

Columns and Expressions

As mentioned previously, named columns in DataFrames are conceptually similar to named columns in pandas or R DataFrames or in an RDBMS table: they describe a type of field. You can list all the columns by their names, and you can perform operations on their values using relational or computational expressions. In Spark's supported languages, columns are objects with public methods (represented by the Column type).

You can also use logical or mathematical expressions on columns. For example, you could create a simple expression using expr("columnName * 5") or (expr("colum nName - 5") > col(anothercolumnName)), where columnName is a Spark type (integer, string, etc.). expr() is part of the pyspark.sql.functions (Python) and org.apache.spark.sql.functions (Scala) packages. Like any other function in those packages, expr() takes arguments that Spark will parse as an expression, computing the result.

 Scala, Java, and Python all have public methods associated with columns (*https://oreil.ly/xVBIX*). You'll note that the Spark documentation refers to both col and Column. Column is the name of the object, while col() is a standard built-in function that returns a Column.

Let's take a look at some examples of what we can do with columns in Spark. Each example is followed by its output:

```scala
// In Scala
scala> import org.apache.spark.sql.functions._
scala> blogsDF.columns
res2: Array[String] = Array(Campaigns, First, Hits, Id, Last, Published, Url)

// Access a particular column with col and it returns a Column type
scala> blogsDF.col("Id")
res3: org.apache.spark.sql.Column = id

// Use an expression to compute a value
scala> blogsDF.select(expr("Hits * 2")).show(2)
// or use col to compute value
scala> blogsDF.select(col("Hits") * 2).show(2)

+----------+
|(Hits * 2)|
+----------+
|      9070|
|     17816|
+----------+

// Use an expression to compute big hitters for blogs
// This adds a new column, Big Hitters, based on the conditional expression
blogsDF.withColumn("Big Hitters", (expr("Hits > 10000"))).show()

+---+---------+-------+---+----------+-----+--------------------+-----------+
| Id|    First|   Last|Url| Published| Hits|           Campaigns|Big Hitters|
+---+---------+-------+---+----------+-----+--------------------+-----------+
|  1|    Jules|  Damji|...|  1/4/2016| 4535| [twitter, LinkedIn]|      false|
|  2|   Brooke|  Wenig|...|  5/5/2018| 8908| [twitter, LinkedIn]|      false|
|  3|    Denny|    Lee|...|  6/7/2019| 7659|[web, twitter, FB...|      false|
|  4|Tathagata|    Das|...|5/12/2018|10568|       [twitter, FB]|       true|
```

```
|  5|    Matei|Zaharia|...|5/14/2014|40578|[web, twitter, FB...|        true|
|  6|  Reynold|    Xin|...| 3/2/2015|25568| [twitter, LinkedIn]|        true|
+---+---------+-------+---+---------+-----+--------------------+-----------+
```

```
// Concatenate three columns, create a new column, and show the
// newly created concatenated column
blogsDF
  .withColumn("AuthorsId", (concat(expr("First"), expr("Last"), expr("Id"))))
  .select(col("AuthorsId"))
  .show(4)
```

```
+-------------+
|    AuthorsId|
+-------------+
|  JulesDamji1|
| BrookeWenig2|
|    DennyLee3|
|TathagataDas4|
+-------------+
```

```
// These statements return the same value, showing that
// expr is the same as a col method call
blogsDF.select(expr("Hits")).show(2)
blogsDF.select(col("Hits")).show(2)
blogsDF.select("Hits").show(2)
```

```
+-----+
| Hits|
+-----+
| 4535|
| 8908|
+-----+
```

```
// Sort by column "Id" in descending order
blogsDF.sort(col("Id").desc).show()
blogsDF.sort($"Id".desc).show()
```

```
+--------------------+---------+-----+---+--------+---------+----------------+
|           Campaigns|    First| Hits| Id|    Last|Published|             Url|
+--------------------+---------+-----+---+--------+---------+----------------+
| [twitter, LinkedIn]|  Reynold|25568|  6|     Xin| 3/2/2015|https://tinyurl.6|
|[web, twitter, FB...|    Matei|40578|  5| Zaharia|5/14/2014|https://tinyurl.5|
|        [twitter, FB]|Tathagata|10568|  4|     Das|5/12/2018|https://tinyurl.4|
|[web, twitter, FB...|    Denny| 7659|  3|     Lee| 6/7/2019|https://tinyurl.3|
| [twitter, LinkedIn]|   Brooke| 8908|  2|   Wenig| 5/5/2018|https://tinyurl.2|
| [twitter, LinkedIn]|    Jules| 4535|  1|   Damji| 1/4/2016|https://tinyurl.1|
+--------------------+---------+-----+---+--------+---------+----------------+
```

In this last example, the expressions blogs_df.sort(col("Id").desc) and blogsDF.sort($"Id".desc) are identical. They both sort the DataFrame column named Id in descending order: one uses an explicit function, col("Id"), to return a

Column object, while the other uses $ before the name of the column, which is a function in Spark that converts column named Id to a Column.

 We have only scratched the surface here, and employed just a couple of methods on Column objects. For a complete list of all public methods for Column objects, we refer you to the Spark documentation (*https://oreil.ly/TZd3c*).

Column objects in a DataFrame can't exist in isolation; each column is part of a row in a record and all the rows together constitute a DataFrame, which as we will see later in the chapter is really a Dataset[Row] in Scala.

Rows

A row in Spark is a generic Row object (*https://oreil.ly/YLMnw*), containing one or more columns. Each column may be of the same data type (e.g., integer or string), or they can have different types (integer, string, map, array, etc.). Because Row is an object in Spark and an ordered collection of fields, you can instantiate a Row in each of Spark's supported languages and access its fields by an index starting at 0:

```
// In Scala
import org.apache.spark.sql.Row
// Create a Row
val blogRow = Row(6, "Reynold", "Xin", "https://tinyurl.6", 255568, "3/2/2015",
  Array("twitter", "LinkedIn"))
// Access using index for individual items
blogRow(1)
res62: Any = Reynold
```

```
# In Python
from pyspark.sql import Row
blog_row = Row(6, "Reynold", "Xin", "https://tinyurl.6", 255568, "3/2/2015",
  ["twitter", "LinkedIn"])
# access using index for individual items
blog_row[1]
'Reynold'
```

Row objects can be used to create DataFrames if you need them for quick interactivity and exploration:

```
# In Python
rows = [Row("Matei Zaharia", "CA"), Row("Reynold Xin", "CA")]
authors_df = spark.createDataFrame(rows, ["Authors", "State"])
authors_df.show()
```

```
// In Scala
val rows = Seq(("Matei Zaharia", "CA"), ("Reynold Xin", "CA"))
val authorsDF = rows.toDF("Author", "State")
authorsDF.show()
```

```
+---------------+-----+
|        Author|State|
+---------------+-----+
|Matei Zaharia|   CA|
|  Reynold Xin|   CA|
+---------------+-----+
```

In practice, though, you will usually want to read DataFrames from a file as illustrated earlier. In most cases, because your files are going to be huge, defining a schema and using it is a quicker and more efficient way to create DataFrames.

After you have created a large distributed DataFrame, you are going to want to perform some common data operations on it. Let's examine some of the Spark operations you can perform with high-level relational operators in the Structured APIs.

Common DataFrame Operations

To perform common data operations on DataFrames, you'll first need to load a Data-Frame from a data source that holds your structured data. Spark provides an interface, DataFrameReader (*https://oreil.ly/v3WLZ*), that enables you to read data into a DataFrame from myriad data sources in formats such as JSON, CSV, Parquet, Text, Avro, ORC, etc. Likewise, to write a DataFrame back to a data source in a particular format, Spark uses DataFrameWriter (*https://oreil.ly/vzjau*).

Using DataFrameReader and DataFrameWriter

Reading and writing are simple in Spark because of these high-level abstractions and contributions from the community to connect to a wide variety of data sources, including common NoSQL stores, RDBMSs, streaming engines such as Apache Kafka and Kinesis, and more.

To get started, let's read a large CSV file containing data on San Francisco Fire Department calls.[1] As noted previously, we will define a schema for this file and use the DataFrameReader class and its methods to tell Spark what to do. Because this file contains 28 columns and over 4,380,660 records,[2] it's more efficient to define a schema than have Spark infer it.

---

1 This public data is available at https://oreil.ly/iDzQK.

2 The original data set has over 60 columns. We dropped a few unnecessary columns, removed records with null or invalid values, and added an extra Delay column.

If you don't want to specify the schema, Spark can infer schema from a sample at a lesser cost. For example, you can use the `samplingRatio` option:

```scala
// In Scala
val sampleDF = spark
  .read
  .option("samplingRatio", 0.001)
  .option("header", true)
  .csv("""/databricks-datasets/learning-spark-v2/
  sf-fire/sf-fire-calls.csv""")
```

Let's take a look at how to do this:

```python
# In Python, define a schema
from pyspark.sql.types import *

# Programmatic way to define a schema
fire_schema = StructType([StructField('CallNumber', IntegerType(), True),
                StructField('UnitID', StringType(), True),
                StructField('IncidentNumber', IntegerType(), True),
                StructField('CallType', StringType(), True),
                StructField('CallDate', StringType(), True),
                StructField('WatchDate', StringType(), True),
                StructField('CallFinalDisposition', StringType(), True),
                StructField('AvailableDtTm', StringType(), True),
                StructField('Address', StringType(), True),
                StructField('City', StringType(), True),
                StructField('Zipcode', IntegerType(), True),
                StructField('Battalion', StringType(), True),
                StructField('StationArea', StringType(), True),
                StructField('Box', StringType(), True),
                StructField('OriginalPriority', StringType(), True),
                StructField('Priority', StringType(), True),
                StructField('FinalPriority', IntegerType(), True),
                StructField('ALSUnit', BooleanType(), True),
                StructField('CallTypeGroup', StringType(), True),
                StructField('NumAlarms', IntegerType(), True),
                StructField('UnitType', StringType(), True),
                StructField('UnitSequenceInCallDispatch', IntegerType(), True),
                StructField('FirePreventionDistrict', StringType(), True),
                StructField('SupervisorDistrict', StringType(), True),
                StructField('Neighborhood', StringType(), True),
                StructField('Location', StringType(), True),
                StructField('RowID', StringType(), True),
                StructField('Delay', FloatType(), True)])

# Use the DataFrameReader interface to read a CSV file
sf_fire_file = "/databricks-datasets/learning-spark-v2/sf-fire/sf-fire-calls.csv"
fire_df = spark.read.csv(sf_fire_file, header=True, schema=fire_schema)
```

```scala
// In Scala it would be similar
val fireSchema = StructType(Array(StructField("CallNumber", IntegerType, true),
```

```
                          StructField("UnitID", StringType, true),
                          StructField("IncidentNumber", IntegerType, true),
                          StructField("CallType", StringType, true),
                          StructField("Location", StringType, true),
                          ...
                          ...
                          StructField("Delay", FloatType, true)))

// Read the file using the CSV DataFrameReader
val sfFireFile="/databricks-datasets/learning-spark-v2/sf-fire/sf-fire-calls.csv"
val fireDF = spark.read.schema(fireSchema)
  .option("header", "true")
  .csv(sfFireFile)
```

The `spark.read.csv()` function reads in the CSV file and returns a DataFrame of rows and named columns with the types dictated in the schema.

To write the DataFrame into an external data source in your format of choice, you can use the `DataFrameWriter` interface. Like `DataFrameReader`, it supports multiple data sources (*https://oreil.ly/4rYNZ*). Parquet, a popular columnar format, is the default format; it uses snappy compression to compress the data. If the DataFrame is written as Parquet, the schema is preserved as part of the Parquet metadata. In this case, subsequent reads back into a DataFrame do not require you to manually supply a schema.

Saving a DataFrame as a Parquet file or SQL table. A common data operation is to explore and transform your data, and then persist the DataFrame in Parquet format or save it as a SQL table. Persisting a transformed DataFrame is as easy as reading it. For example, to persist the DataFrame we were just working with as a file after reading it you would do the following:

```
// In Scala to save as a Parquet file
val parquetPath = ...
fireDF.write.format("parquet").save(parquetPath)

# In Python to save as a Parquet file
parquet_path = ...
fire_df.write.format("parquet").save(parquet_path)
```

Alternatively, you can save it as a table, which registers metadata with the Hive metastore (we will cover SQL managed and unmanaged tables, metastores, and DataFrames in the next chapter):

```
// In Scala to save as a table
val parquetTable = ... // name of the table
fireDF.write.format("parquet").saveAsTable(parquetTable)

# In Python
parquet_table = ... # name of the table
fire_df.write.format("parquet").saveAsTable(parquet_table)
```

Let's walk through some common operations to perform on DataFrames after you have read the data.

Transformations and actions

Now that you have a distributed DataFrame composed of San Francisco Fire Department calls in memory, the first thing you as a developer will want to do is examine your data to see what the columns look like. Are they of the correct types? Do any of them need to be converted to different types? Do they have null values?

In "Transformations, Actions, and Lazy Evaluation" on page 28 in Chapter 2, you got a glimpse of how transformations and actions are used to operate on DataFrames, and saw some common examples of each. What can we find out from our San Francisco Fire Department calls using these?

Projections and filters. A *projection* in relational parlance is a way to return only the rows matching a certain relational condition by using filters. In Spark, projections are done with the select() method, while filters can be expressed using the filter() or where() method. We can use this technique to examine specific aspects of our SF Fire Department data set:

```python
# In Python
few_fire_df = (fire_df
  .select("IncidentNumber", "AvailableDtTm", "CallType")
  .where(col("CallType") != "Medical Incident"))
few_fire_df.show(5, truncate=False)
```

```scala
// In Scala
val fewFireDF = fireDF
  .select("IncidentNumber", "AvailableDtTm", "CallType")
  .where($"CallType" =!= "Medical Incident")
fewFireDF.show(5, false)
```

```
+--------------+----------------------+--------------+
|IncidentNumber|AvailableDtTm         |CallType      |
+--------------+----------------------+--------------+
|2003235       |01/11/2002 01:47:00 AM|Structure Fire|
|2003235       |01/11/2002 01:51:54 AM|Structure Fire|
|2003235       |01/11/2002 01:47:00 AM|Structure Fire|
|2003235       |01/11/2002 01:47:00 AM|Structure Fire|
|2003235       |01/11/2002 01:51:17 AM|Structure Fire|
+--------------+----------------------+--------------+
only showing top 5 rows
```

What if we want to know how many distinct CallTypes were recorded as the causes of the fire calls? These simple and expressive queries do the job:

```python
# In Python, return number of distinct types of calls using countDistinct()
from pyspark.sql.functions import *
(fire_df
```

```
  .select("CallType")
  .where(col("CallType").isNotNull())
  .agg(countDistinct("CallType").alias("DistinctCallTypes"))
  .show())

// In Scala
import org.apache.spark.sql.functions._
fireDF
  .select("CallType")
  .where(col("CallType").isNotNull)
  .agg(countDistinct('CallType) as 'DistinctCallTypes)
  .show()
```

```
+-----------------+
|DistinctCallTypes|
+-----------------+
|               32|
+-----------------+
```

We can list the distinct call types in the data set using these queries:

```
# In Python, filter for only distinct non-null CallTypes from all the rows
(fire_df
  .select("CallType")
  .where(col("CallType").isNotNull())
  .distinct()
  .show(10, False))

// In Scala
fireDF
  .select("CallType")
  .where($"CallType".isNotNull())
  .distinct()
  .show(10, false)
```

```
Out[20]: 32
```

```
+--------------------------------+
|CallType                        |
+--------------------------------+
|Elevator / Escalator Rescue     |
|Marine Fire                     |
|Aircraft Emergency              |
|Confined Space / Structure Collapse|
|Administrative                  |
|Alarms                          |
|Odor (Strange / Unknown)        |
|Lightning Strike (Investigation)|
|Citizen Assist / Service Call   |
|HazMat                          |
+--------------------------------+
only showing top 10 rows
```

Renaming, adding, and dropping columns. Sometimes you want to rename particular columns for reasons of style or convention, and at other times for readability or brevity. The original column names in the SF Fire Department data set had spaces in them. For example, the column name `IncidentNumber` was `Incident Number`. Spaces in column names can be problematic, especially when you want to write or save a DataFrame as a Parquet file (which prohibits this).

By specifying the desired column names in the schema with `StructField`, as we did, we effectively changed all names in the resulting DataFrame.

Alternatively, you could selectively rename columns with the `withColumnRenamed()` method. For instance, let's change the name of our `Delay` column to `ResponseDe layedinMins` and take a look at the response times that were longer than five minutes:

```python
# In Python
new_fire_df = fire_df.withColumnRenamed("Delay", "ResponseDelayedinMins")
(new_fire_df
  .select("ResponseDelayedinMins")
  .where(col("ResponseDelayedinMins") > 5)
  .show(5, False))
```

```scala
// In Scala
val newFireDF = fireDF.withColumnRenamed("Delay", "ResponseDelayedinMins")
newFireDF
  .select("ResponseDelayedinMins")
  .where($"ResponseDelayedinMins" > 5)
  .show(5, false)
```

This gives us a new renamed column:

```
+---------------------+
|ResponseDelayedinMins|
+---------------------+
|5.233333             |
|6.9333334            |
|6.116667             |
|7.85                 |
|77.333336            |
+---------------------+
only showing top 5 rows
```

 Because DataFrame transformations are immutable, when we rename a column using `withColumnRenamed()` we get a new Data-Frame while retaining the original with the old column name.

Modifying the contents of a column or its type are common operations during data exploration. In some cases the data is raw or dirty, or its types are not amenable to being supplied as arguments to relational operators. For example, in our SF Fire Department data set, the columns `CallDate`, `WatchDate`, and `AlarmDtTm` are strings rather than either Unix timestamps or SQL dates, both of which Spark supports and can easily manipulate during transformations or actions (e.g., during a date- or time-based analysis of the data).

So how do we convert them into a more usable format? It's quite simple, thanks to some high-level API methods. `spark.sql.functions` has a set of to/from date/timestamp functions such as `to_timestamp()` and `to_date()` that we can use for just this purpose:

```Python
# In Python
fire_ts_df = (new_fire_df
  .withColumn("IncidentDate", to_timestamp(col("CallDate"), "MM/dd/yyyy"))
  .drop("CallDate")
  .withColumn("OnWatchDate", to_timestamp(col("WatchDate"), "MM/dd/yyyy"))
  .drop("WatchDate")
  .withColumn("AvailableDtTS", to_timestamp(col("AvailableDtTm"),
  "MM/dd/yyyy hh:mm:ss a"))
  .drop("AvailableDtTm"))

# Select the converted columns
(fire_ts_df
  .select("IncidentDate", "OnWatchDate", "AvailableDtTS")
  .show(5, False))
```
```Scala
// In Scala
val fireTsDF = newFireDF
  .withColumn("IncidentDate", to_timestamp(col("CallDate"), "MM/dd/yyyy"))
  .drop("CallDate")
  .withColumn("OnWatchDate", to_timestamp(col("WatchDate"), "MM/dd/yyyy"))
  .drop("WatchDate")
  .withColumn("AvailableDtTS", to_timestamp(col("AvailableDtTm"),
  "MM/dd/yyyy hh:mm:ss a"))
  .drop("AvailableDtTm")

// Select the converted columns
fireTsDF
  .select("IncidentDate", "OnWatchDate", "AvailableDtTS")
  .show(5, false)
```

Those queries pack quite a punch—a number of things are happening. Let's unpack what they do:

1. Convert the existing column's data type from string to a Spark-supported timestamp.

2. Use the new format specified in the format string "MM/dd/yyyy" or "MM/dd/yyyy hh:mm:ss a" where appropriate.

3. After converting to the new data type, drop() the old column and append the new one specified in the first argument to the withColumn() method.

4. Assign the new modified DataFrame to fire_ts_df.

The queries result in three new columns:

```
+-------------------+-------------------+-------------------+
|IncidentDate       |OnWatchDate        |AvailableDtTS      |
+-------------------+-------------------+-------------------+
|2002-01-11 00:00:00|2002-01-10 00:00:00|2002-01-11 01:58:43|
|2002-01-11 00:00:00|2002-01-10 00:00:00|2002-01-11 02:10:17|
|2002-01-11 00:00:00|2002-01-10 00:00:00|2002-01-11 01:47:00|
|2002-01-11 00:00:00|2002-01-10 00:00:00|2002-01-11 01:51:54|
|2002-01-11 00:00:00|2002-01-10 00:00:00|2002-01-11 01:47:00|
+-------------------+-------------------+-------------------+
only showing top 5 rows
```

Now that we have modified the dates, we can query using functions from spark.sql.functions like dayofmonth(), dayofyear(), and dayofweek() to explore our data further. We could find out how many calls were logged in the last seven days, or we could see how many years' worth of Fire Department calls are included in the data set with this query:

```python
# In Python
(fire_ts_df
  .select(year('IncidentDate'))
  .distinct()
  .orderBy(year('IncidentDate'))
  .show())
```

```scala
// In Scala
fireTsDF
  .select(year($"IncidentDate"))
  .distinct()
  .orderBy(year($"IncidentDate"))
  .show()
```

```
+------------------+
|year(IncidentDate)|
+------------------+
|              2000|
|              2001|
|              2002|
|              2003|
|              2004|
|              2005|
```

```
|              2006|
|              2007|
|              2008|
|              2009|
|              2010|
|              2011|
|              2012|
|              2013|
|              2014|
|              2015|
|              2016|
|              2017|
|              2018|
+------------------+
```

So far in this section, we have explored a number of common data operations: reading and writing DataFrames; defining a schema and using it when reading in a DataFrame; saving a DataFrame as a Parquet file or table; projecting and filtering selected columns from an existing DataFrame; and modifying, renaming, and dropping columns.

One final common operation is grouping data by values in a column and aggregating the data in some way, like simply counting it. This pattern of grouping and counting is as common as projecting and filtering. Let's have a go at it.

Aggregations.   What if we want to know what the most common types of fire calls were, or what zip codes accounted for the most calls? These kinds of questions are common in data analysis and exploration.

A handful of transformations and actions on DataFrames, such as `groupBy()`, `orderBy()`, and `count()`, offer the ability to aggregate by column names and then aggregate counts across them.

 For larger DataFrames on which you plan to conduct frequent or repeated queries, you could benefit from caching. We will cover DataFrame caching strategies and their benefits in later chapters.

Let's take our first question: what were the most common types of fire calls?

```python
# In Python
(fire_ts_df
  .select("CallType")
  .where(col("CallType").isNotNull())
  .groupBy("CallType")
  .count()
  .orderBy("count", ascending=False)
  .show(n=10, truncate=False))
```

```scala
// In Scala
fireTsDF
  .select("CallType")
  .where(col("CallType").isNotNull)
  .groupBy("CallType")
  .count()
  .orderBy(desc("count"))
  .show(10, false)
```

```
+--------------------------------+-------+
|CallType                        |count  |
+--------------------------------+-------+
|Medical Incident                |2843475|
|Structure Fire                  |578998 |
|Alarms                          |483518 |
|Traffic Collision               |175507 |
|Citizen Assist / Service Call   |65360  |
|Other                           |56961  |
|Outside Fire                    |51603  |
|Vehicle Fire                    |20939  |
|Water Rescue                    |20037  |
|Gas Leak (Natural and LP Gases) |17284  |
+--------------------------------+-------+
```

From this output we can conclude that the most common call type is Medical Incident.

 The DataFrame API also offers the collect() method, but for extremely large DataFrames this is resource-heavy (expensive) and dangerous, as it can cause out-of-memory (OOM) exceptions. Unlike count(), which returns a single number to the driver, collect() returns a collection of all the Row objects in the entire DataFrame or Dataset. If you want to take a peek at some Row records you're better off with take(n), which will return only the first n Row objects of the DataFrame.

Other common DataFrame operations. Along with all the others we've seen, the DataFrame API provides descriptive statistical methods like min(), max(), sum(), and avg(). Let's take a look at some examples showing how to compute them with our SF Fire Department data set.

Here we compute the sum of alarms, the average response time, and the minimum and maximum response times to all fire calls in our data set, importing the PySpark functions in a Pythonic way so as not to conflict with the built-in Python functions:

```python
# In Python
import pyspark.sql.functions as F
(fire_ts_df
  .select(F.sum("NumAlarms"), F.avg("ResponseDelayedinMins"),
```

```
    F.min("ResponseDelayedinMins"), F.max("ResponseDelayedinMins"))
  .show())

// In Scala
import org.apache.spark.sql.{functions => F}
fireTsDF
  .select(F.sum("NumAlarms"), F.avg("ResponseDelayedinMins"),
  F.min("ResponseDelayedinMins"), F.max("ResponseDelayedinMins"))
  .show()
```

```
+--------------+------------------------+------------------------+--------+
|sum(NumAlarms)|avg(ResponseDelayedinMins)|min(ResponseDelayedinMins)|max(...)|
+--------------+------------------------+------------------------+--------+
|       4403441|        3.902170335891614|         0.016666668|1879.6167|
+--------------+------------------------+------------------------+--------+
```

For more advanced statistical needs common with data science workloads, read the API documentation for methods like stat(), describe(), correlation(), covariance(), sampleBy(), approxQuantile(), frequentItems(), and so on.

As you can see, it's easy to compose and chain expressive queries with DataFrames' high-level API and DSL operators. We can't imagine the opacity and comparative unreadability of the code if we were to try to do the same with RDDs!

End-to-End DataFrame Example

There are many possibilities for exploratory data analysis, ETL, and common data operations on the San Francisco Fire Department public data set, above and beyond what we've shown here.

For brevity we won't include all the example code here, but the book's GitHub repo (*https://github.com/databricks/LearningSparkV2*) provides Python and Scala notebooks for you to try to complete an end-to-end DataFrame example using this data set. The notebooks explore and answer the following common questions that you might ask, using the DataFrame API and DSL relational operators:

- What were all the different types of fire calls in 2018?
- What months within the year 2018 saw the highest number of fire calls?
- Which neighborhood in San Francisco generated the most fire calls in 2018?
- Which neighborhoods had the worst response times to fire calls in 2018?
- Which week in the year in 2018 had the most fire calls?
- Is there a correlation between neighborhood, zip code, and number of fire calls?
- How can we use Parquet files or SQL tables to store this data and read it back?

So far we have extensively discussed the DataFrame API, one of the Structured APIs that span Spark's MLlib and Structured Streaming components, which we cover later in the book.

Next, we'll shift our focus to the Dataset API and explore how the two APIs provide a unified, structured interface to developers for programming Spark. We'll then examine the relationship between the RDD, DataFrame, and Dataset APIs, and help you determine when to use which API and why.

The Dataset API

As stated earlier in this chapter, Spark 2.0 unified (*https://oreil.ly/t3RGF*) the DataFrame and Dataset APIs as Structured APIs with similar interfaces so that developers would only have to learn a single set of APIs. Datasets take on two characteristics: *typed* and *untyped* APIs (*https://oreil.ly/_3quT*), as shown in Figure 3-1.

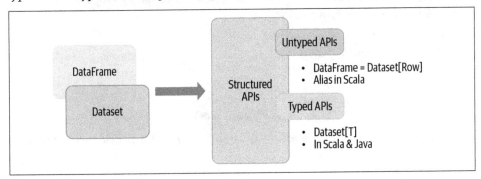

Figure 3-1. Structured APIs in Apache Spark

Conceptually, you can think of a DataFrame in Scala as an alias for a collection of generic objects, `Dataset[Row]`, where a `Row` is a generic untyped JVM object that may hold different types of fields. A Dataset, by contrast, is a collection of strongly typed JVM objects in Scala or a class in Java. Or, as the Dataset documentation (*https://oreil.ly/wSkcJ*) puts it, a Dataset is:

> a strongly typed collection of domain-specific objects that can be transformed in parallel using functional or relational operations. Each Dataset [in Scala] also has an untyped view called a DataFrame, which is a Dataset of Row.

Typed Objects, Untyped Objects, and Generic Rows

In Spark's supported languages, Datasets make sense only in Java and Scala, whereas in Python and R only DataFrames make sense. This is because Python and R are not compile-time type-safe; types are dynamically inferred or assigned during execution, not during compile time. The reverse is true in Scala and Java: types are bound to variables and objects at compile time. In Scala, however, a DataFrame is just an alias for untyped `Dataset[Row]`. Table 3-6 distills it in a nutshell.

Table 3-6. Typed and untyped objects in Spark

Language	Typed and untyped main abstraction	Typed or untyped
Scala	`Dataset[T]` and DataFrame (alias for `Dataset[Row]`)	Both typed and untyped
Java	`Dataset<T>`	Typed
Python	DataFrame	Generic Row untyped
R	DataFrame	Generic Row untyped

Row is a generic object type in Spark, holding a collection of mixed types that can be accessed using an index. Internally, Spark manipulates Row objects, converting them to the equivalent types covered in Table 3-2 and Table 3-3. For example, an Int as one of your fields in a Row will be mapped or converted to IntegerType or IntegerType() respectively for Scala or Java and Python:

```
// In Scala
import org.apache.spark.sql.Row
val row = Row(350, true, "Learning Spark 2E", null)

# In Python
from pyspark.sql import Row
row = Row(350, True, "Learning Spark 2E", None)
```

Using an index into the Row object, you can access individual fields with its public *getter* methods:

```
// In Scala
row.getInt(0)
res23: Int = 350
row.getBoolean(1)
res24: Boolean = true
row.getString(2)
res25: String = Learning Spark 2E

# In Python
row[0]
Out[13]: 350
row[1]
Out[14]: True
row[2]
Out[15]: 'Learning Spark 2E'
```

By contrast, typed objects are actual Java or Scala class objects in the JVM. Each element in a Dataset maps to a JVM object.

Creating Datasets

As with creating DataFrames from data sources, when creating a Dataset you have to know the schema. In other words, you need to know the data types. Although with JSON and CSV data it's possible to infer the schema, for large data sets this is resource-intensive (expensive). When creating a Dataset in Scala, the easiest way to specify the schema for the resulting Dataset is to use a case class. In Java, JavaBean classes are used (we further discuss JavaBean and Scala case class in Chapter 6).

Scala: Case classes

When you wish to instantiate your own domain-specific object as a Dataset, you can do so by defining a case class in Scala. As an example, let's look at a collection of readings from Internet of Things (IoT) devices in a JSON file (we use this file in the end-to-end example later in this section).

Our file has rows of JSON strings that look as follows:

```
{"device_id": 198164, "device_name": "sensor-pad-198164owomcJZ", "ip":
"80.55.20.25", "cca2": "PL", "cca3": "POL", "cn": "Poland", "latitude":
53.080000, "longitude": 18.620000, "scale": "Celsius", "temp": 21,
"humidity": 65, "battery_level": 8, "c02_level": 1408,"lcd": "red",
"timestamp" :1458081226051}
```

To express each JSON entry as `DeviceIoTData`, a domain-specific object, we can define a Scala case class:

```
case class DeviceIoTData (battery_level: Long, c02_level: Long,
cca2: String, cca3: String, cn: String, device_id: Long,
device_name: String, humidity: Long, ip: String, latitude: Double,
lcd: String, longitude: Double, scale:String, temp: Long,
timestamp: Long)
```

Once defined, we can use it to read our file and convert the returned `Dataset[Row]` into `Dataset[DeviceIoTData]` (output truncated to fit on the page):

```
// In Scala
val ds = spark.read
 .json("/databricks-datasets/learning-spark-v2/iot-devices/iot_devices.json")
 .as[DeviceIoTData]

ds: org.apache.spark.sql.Dataset[DeviceIoTData] = [battery_level...]

ds.show(5, false)
```

```
+-------------+---------+----+----+-------------+---------+---+
|battery_level|c02_level|cca2|cca3|cn           |device_id|...|
+-------------+---------+----+----+-------------+---------+---+
|8            |868      |US  |USA |United States|1        |...|
|7            |1473     |NO  |NOR |Norway       |2        |...|
|2            |1556     |IT  |ITA |Italy        |3        |...|
```

```
|6             |1080     |US  |USA |United States|4         |...|
|4             |931      |PH  |PHL |Philippines  |5         |...|
+--------------|---------|----|----|-------------|----------|---+
only showing top 5 rows
```

Dataset Operations

Just as you can perform transformations and actions on DataFrames, so you can with Datasets. Depending on the kind of operation, the results will vary:

```scala
// In Scala
val filterTempDS = ds.filter({d => {d.temp > 30 && d.humidity > 70}})

filterTempDS: org.apache.spark.sql.Dataset[DeviceIoTData] = [battery_level...]

filterTempDS.show(5, false)
```

```
+--------------|---------|----|----|-------------|----------|---+
|battery_level|c02_level|cca2|cca3|cn           |device_id|...|
+--------------|---------|----|----|-------------|----------|---+
|0             |1466     |US  |USA |United States|17        |...|
|9             |986      |FR  |FRA |France       |48        |...|
|8             |1436     |US  |USA |United States|54        |...|
|4             |1090     |US  |USA |United States|63        |...|
|4             |1072     |PH  |PHL |Philippines  |81        |...|
+--------------|---------|----|----|-------------|----------|---+
only showing top 5 rows
```

In this query, we used a function as an argument to the Dataset method `filter()`. This is an overloaded method with many signatures. The version we used, `filter(func: (T) > Boolean): Dataset[T]`, takes a lambda function, `func: (T) > Boolean`, as its argument.

The argument to the lambda function is a JVM object of type `DeviceIoTData`. As such, we can access its individual data fields using the dot (.) notation, like you would in a Scala class or JavaBean.

Another thing to note is that with DataFrames, you express your `filter()` conditions as SQL-like DSL operations, which are language-agnostic (as we saw earlier in the fire calls examples). With Datasets, we use language-native expressions as Scala or Java code.

Here's another example that results in another, smaller Dataset:

```scala
// In Scala
case class DeviceTempByCountry(temp: Long, device_name: String, device_id: Long,
  cca3: String)
val dsTemp = ds
  .filter(d => {d.temp > 25})
  .map(d => (d.temp, d.device_name, d.device_id, d.cca3))
  .toDF("temp", "device_name", "device_id", "cca3")
```

```
  .as[DeviceTempByCountry]
dsTemp.show(5, false)
```

```
+----+--------------------+---------+----+
|temp|device_name         |device_id|cca3|
+----+--------------------+---------+----+
|34  |meter-gauge-1xbYRYcj|1        |USA |
|28  |sensor-pad-4mzWkz    |4        |USA |
|27  |sensor-pad-6al7RTAobR|6       |USA |
|27  |sensor-pad-8xUD6pzsQI|8       |JPN |
|26  |sensor-pad-10BsywSYUF|10      |USA |
+----+--------------------+---------+----+
only showing top 5 rows
```

Or you can inspect only the first row of your Dataset:

```
val device = dsTemp.first()
println(device)

device: DeviceTempByCountry =
DeviceTempByCountry(34,meter-gauge-1xbYRYcj,1,USA)
```

Alternatively, you could express the same query using column names and then cast to a `Dataset[DeviceTempByCountry]`:

```
// In Scala
val dsTemp2 = ds
  .select($"temp", $"device_name", $"device_id", $"device_id", $"cca3")
  .where("temp > 25")
  .as[DeviceTempByCountry]
```

> Semantically, `select()` is like `map()` in the previous query, in that both of these queries select fields and generate equivalent results.

To recap, the operations we can perform on Datasets—`filter()`, `map()`, `groupBy()`, `select()`, `take()`, etc.—are similar to the ones on DataFrames. In a way, Datasets are similar to RDDs in that they provide a similar interface to its aforementioned methods and compile-time safety but with a much easier to read and an object-oriented programming interface.

When we use Datasets, the underlying Spark SQL engine handles the creation, conversion, serialization, and deserialization of the JVM objects. It also takes care of off-Java heap memory management with the help of Dataset encoders. (We will talk more about Datasets and memory management in Chapter 6.)

End-to-End Dataset Example

In this end-to-end Dataset example you'll conduct similar exploratory data analysis, ETL (extract, transform, and load), and data operations as in the DataFrame example, using the IoT data set. This data set is small and fake, but our main goal here is to illustrate the clarity with which you can express queries with Datasets and the readability of those queries, just as we did with DataFrames.

Again, for brevity, we won't include all the example code here; however, we have furnished the notebook in the GitHub repo (*https://github.com/databricks/Learning SparkV2/*). The notebook explores common operations you might conduct with this data set. Using the Dataset API, we attempt to do the following:

1. Detect failing devices with battery levels below a threshold.

2. Identify offending countries with high levels of CO2 emissions.

3. Compute the min and max values for temperature, battery level, CO2, and humidity.

4. Sort and group by average temperature, CO2, humidity, and country.

DataFrames Versus Datasets

By now you may be wondering why and when you should use DataFrames or Datasets. In many cases either will do, depending on the languages you are working in, but there are some situations where one is preferable to the other. Here are a few examples:

- If you want to tell Spark *what to do*, not *how to do it*, use DataFrames or Datasets.

- If you want rich semantics, high-level abstractions, and DSL operators, use DataFrames or Datasets.

- If you want strict compile-time type safety and don't mind creating multiple case classes for a specific `Dataset[T]`, use Datasets.

- If your processing demands high-level expressions, filters, maps, aggregations, computing averages or sums, SQL queries, columnar access, or use of relational operators on semi-structured data, use DataFrames or Datasets.

- If your processing dictates relational transformations similar to SQL-like queries, use DataFrames.

- If you want to take advantage of and benefit from Tungsten's efficient serialization with Encoders, use Datasets (*https://oreil.ly/13XHQ*).

- If you want unification, code optimization, and simplification of APIs across Spark components, use DataFrames.

- If you are an R user, use DataFrames.
- If you are a Python user, use DataFrames and drop down to RDDs if you need more control.
- If you want space and speed efficiency, use DataFrames.
- If you want errors caught during compilation rather than at runtime, choose the appropriate API as depicted in Figure 3-2.

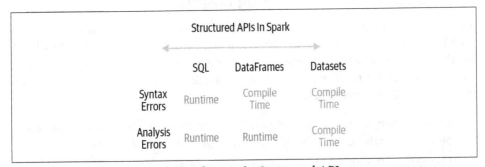

Figure 3-2. When errors are detected using the Structured APIs

When to Use RDDs

You may ask: Are RDDs being relegated to second-class citizens? Are they being deprecated? The answer is a resounding *no*! The RDD API will continue to be supported, although all future development work in Spark 2.x and Spark 3.0 will continue to have a DataFrame interface and semantics rather than using RDDs.

There are some scenarios where you'll want to consider using RDDs, such as when you:

- Are using a third-party package that's written using RDDs
- Can forgo the code optimization, efficient space utilization, and performance benefits available with DataFrames and Datasets
- Want to precisely instruct Spark *how to do* a query

What's more, you can seamlessly move between DataFrames or Datasets and RDDs at will using a simple API method call, df.rdd. (Note, however, that this does have a cost and should be avoided unless necessary.) After all, DataFrames and Datasets are built on top of RDDs, and they get decomposed to compact RDD code during whole-stage code generation, which we discuss in the next section.

Finally, the preceding sections provided some intuition on how Structured APIs in Spark enable developers to use easy and friendly APIs to compose expressive queries on structured data. In other words, you tell Spark *what to do*, not *how to do it*, using high-level operations, and it ascertains the most efficient way to build a query and generates compact code for you.

This process of building efficient queries and generating compact code is the job of the Spark SQL engine. It's the substrate upon which the Structured APIs we've been looking at are built. Let's peek under the hood at that engine now.

Spark SQL and the Underlying Engine

At a programmatic level, Spark SQL allows developers to issue ANSI SQL:2003–compatible queries on structured data with a schema. Since its introduction in Spark 1.3, Spark SQL has evolved into a substantial engine upon which many high-level structured functionalities have been built. Apart from allowing you to issue SQL-like queries on your data, the Spark SQL engine:

- Unifies Spark components and permits abstraction to DataFrames/Datasets in Java, Scala, Python, and R, which simplifies working with structured data sets.
- Connects to the Apache Hive metastore and tables.
- Reads and writes structured data with a specific schema from structured file formats (JSON, CSV, Text, Avro, Parquet, ORC, etc.) and converts data into temporary tables.
- Offers an interactive Spark SQL shell for quick data exploration.
- Provides a bridge to (and from) external tools via standard database JDBC/ODBC connectors.
- Generates optimized query plans and compact code for the JVM, for final execution.

Figure 3-3 shows the components that Spark SQL interacts with to achieve all of this.

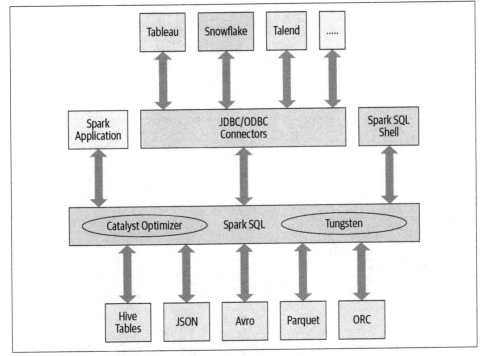

Figure 3-3. Spark SQL and its stack

At the core of the Spark SQL engine are the Catalyst optimizer and Project Tungsten. Together, these support the high-level DataFrame and Dataset APIs and SQL queries. We'll talk more about Tungsten in Chapter 6; for now, let's take a closer look at the optimizer.

The Catalyst Optimizer

The Catalyst optimizer takes a computational query and converts it into an execution plan. It goes through four transformational phases (*https://oreil.ly/jMDOi*), as shown in Figure 3-4:

1. Analysis
2. Logical optimization
3. Physical planning
4. Code generation

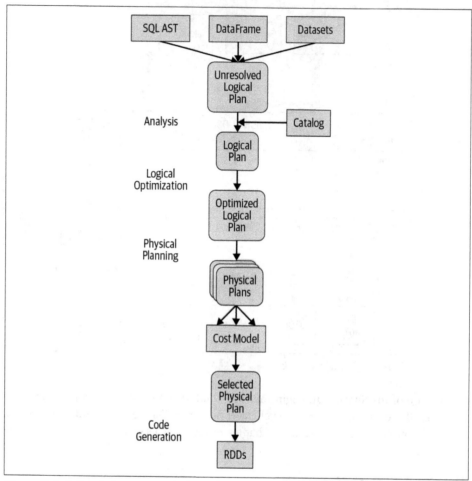

Figure 3-4. A Spark computation's four-phase journey

For example, consider one of the queries from our M&Ms example in Chapter 2. Both of the following sample code blocks will go through the same process, eventually ending up with a similar query plan and identical bytecode for execution. That is, regardless of the language you use, your computation undergoes the same journey and the resulting bytecode is likely the same:

```
# In Python
count_mnm_df = (mnm_df
    .select("State", "Color", "Count")
    .groupBy("State", "Color")
    .agg(sum("Count")
    .alias("Total"))
    .orderBy("Total", ascending=False))
```

```
-- In SQL
SELECT State, Color, sum(Count) AS Total
FROM MNM_TABLE_NAME
GROUP BY State, Color
ORDER BY Total DESC
```

To see the different stages the Python code goes through, you can use the count_mnm_df.explain(True) method on the DataFrame. Or, to get a look at the different logical and physical plans, in Scala you can call df.queryExecution.logical or df.queryExecution.optimizedPlan. (In Chapter 7, we will discuss more about tuning and debugging Spark and how to read query plans.) This gives us the following output:

count_mnm_df.explain(True)

```
== Parsed Logical Plan ==
'Sort ['Total DESC NULLS LAST], true
+- Aggregate [State#10, Color#11], [State#10, Color#11, sum(Count#12) AS...]
   +- Project [State#10, Color#11, Count#12]
      +- Relation[State#10,Color#11,Count#12] csv

== Analyzed Logical Plan ==
State: string, Color: string, Total: bigint
Sort [Total#24L DESC NULLS LAST], true
+- Aggregate [State#10, Color#11], [State#10, Color#11, sum(Count#12) AS...]
   +- Project [State#10, Color#11, Count#12]
      +- Relation[State#10,Color#11,Count#12] csv

== Optimized Logical Plan ==
Sort [Total#24L DESC NULLS LAST], true
+- Aggregate [State#10, Color#11], [State#10, Color#11, sum(Count#12) AS...]
   +- Relation[State#10,Color#11,Count#12] csv

== Physical Plan ==
*(3) Sort [Total#24L DESC NULLS LAST], true, 0
+- Exchange rangepartitioning(Total#24L DESC NULLS LAST, 200)
   +- *(2) HashAggregate(keys=[State#10, Color#11], functions=[sum(Count#12)],
output=[State#10, Color#11, Total#24L])
      +- Exchange hashpartitioning(State#10, Color#11, 200)
         +- *(1) HashAggregate(keys=[State#10, Color#11],
functions=[partial_sum(Count#12)], output=[State#10, Color#11, count#29L])
            +- *(1) FileScan csv [State#10,Color#11,Count#12] Batched: false,
Format: CSV, Location:
InMemoryFileIndex[file:/Users/jules/gits/LearningSpark2.0/chapter2/py/src/...
dataset.csv], PartitionFilters: [], PushedFilters: [], ReadSchema:
struct<State:string,Color:string,Count:int>
```

Let's consider another DataFrame computation example. The following Scala code undergoes a similar journey as the underlying engine optimizes its logical and physical plans:

```
// In Scala
// Users DataFrame read from a Parquet table
val usersDF  = ...
// Events DataFrame read from a Parquet table
val eventsDF = ...
// Join two DataFrames
val joinedDF = users
  .join(events, users("id") === events("uid"))
  .filter(events("date") > "2015-01-01")
```

After going through an initial analysis phase, the query plan is transformed and rearranged by the Catalyst optimizer as shown in Figure 3-5.

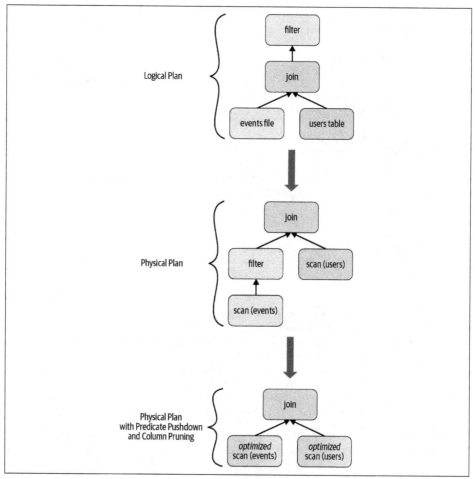

Figure 3-5. An example of a specific query transformation

Let's go through each of the four query optimization phases.

Phase 1: Analysis

The Spark SQL engine begins by generating an abstract syntax tree (AST) (*https://oreil.ly/mOIv4*) for the SQL or DataFrame query. In this initial phase, any columns or table names will be resolved by consulting an internal `Catalog`, a programmatic interface to Spark SQL that holds a list of names of columns, data types, functions, tables, databases, etc. Once they've all been successfully resolved, the query proceeds to the next phase.

Phase 2: Logical optimization

As Figure 3-4 shows, this phase comprises two internal stages. Applying a standard-rule based optimization approach, the Catalyst optimizer will first construct a set of multiple plans and then, using its cost-based optimizer (CBO) (*https://oreil.ly/xVVpP*), assign costs to each plan. These plans are laid out as operator trees (like in Figure 3-5); they may include, for example, the process of constant folding, predicate pushdown, projection pruning, Boolean expression simplification, etc. This logical plan is the input into the physical plan.

Phase 3: Physical planning

In this phase, Spark SQL generates an optimal physical plan for the selected logical plan, using physical operators that match those available in the Spark execution engine.

Phase 4: Code generation

The final phase of query optimization involves generating efficient Java bytecode to run on each machine. Because Spark SQL can operate on data sets loaded in memory, Spark can use state-of-the-art compiler technology for code generation to speed up execution. In other words, it acts as a compiler (*https://oreil.ly/_PYnW*). Project Tungsten, which facilitates whole-stage code generation, plays a role here.

Just what is whole-stage code generation? It's a physical query optimization phase that collapses the whole query into a single function, getting rid of virtual function calls and employing CPU registers for intermediate data. The second-generation Tungsten engine, introduced in Spark 2.0, uses this approach to generate compact RDD code for final execution. This streamlined strategy significantly improves CPU efficiency and performance (*https://oreil.ly/B3A7y*).

We have talked at a conceptual level about the workings of the Spark SQL engine, with its two principal components: the Catalyst optimizer and Project Tungsten. The internal technical workings are beyond the scope of this book; however, for the curious, we encourage you to check out the references in the text for in-depth technical discussions.

Summary

In this chapter, we took a deep dive into Spark's Structured APIs, beginning with a look at the history and merits of structure in Spark.

Through illustrative common data operations and code examples, we demonstrated that the high-level DataFrame and Dataset APIs are far more expressive and intuitive than the low-level RDD API. Designed to make processing of large data sets easier, the Structured APIs provide domain-specific operators for common data operations, increasing the clarity and expressiveness of your code.

We explored when to use RDDs, DataFrames, and Datasets, depending on your use case scenarios.

And finally, we took a look under the hood to see how the Spark SQL engine's main components—the Catalyst optimizer and Project Tungsten—support structured high-level APIs and DSL operators. As you saw, no matter which of the Spark-supported languages you use, a Spark query undergoes the same optimization journey, from logical and physical plan construction to final compact code generation.

The concepts and code examples in this chapter have laid the groundwork for the next two chapters, in which we will further illustrate the seamless interoperability between DataFrames, Datasets, and Spark SQL.

## CHAPTER 4

Spark SQL and DataFrames: Introduction to Built-in Data Sources

In the previous chapter, we explained the evolution of and justification for structure in Spark. In particular, we discussed how the Spark SQL engine provides a unified foundation for the high-level DataFrame and Dataset APIs. Now, we'll continue our discussion of the DataFrame and explore its interoperability with Spark SQL.

This chapter and the next also explore how Spark SQL interfaces with some of the external components shown in Figure 4-1.

In particular, Spark SQL:

- Provides the engine upon which the high-level Structured APIs we explored in Chapter 3 are built.
- Can read and write data in a variety of structured formats (e.g., JSON, Hive tables, Parquet, Avro, ORC, CSV).
- Lets you query data using JDBC/ODBC connectors from external business intelligence (BI) data sources such as Tableau, Power BI, Talend, or from RDBMSs such as MySQL and PostgreSQL.
- Provides a programmatic interface to interact with structured data stored as tables or views in a database from a Spark application
- Offers an interactive shell to issue SQL queries on your structured data.
- Supports ANSI SQL:2003 (*https://oreil.ly/83QYa*)-compliant commands and HiveQL (*https://oreil.ly/QFza4*).

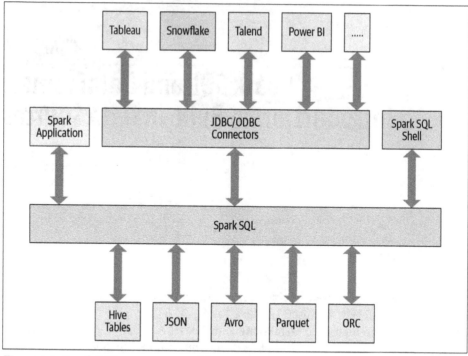

Figure 4-1. Spark SQL connectors and data sources

Let's begin with how you can use Spark SQL in a Spark application.

Using Spark SQL in Spark Applications

The SparkSession, introduced in Spark 2.0, provides a unified entry point (*https:// oreil.ly/B7FZh*) for programming Spark with the Structured APIs. You can use a SparkSession to access Spark functionality: just import the class and create an instance in your code.

To issue any SQL query, use the sql() method on the SparkSession instance, spark, such as spark.sql("SELECT * FROM myTableName"). All spark.sql queries executed in this manner return a DataFrame on which you may perform further Spark operations if you desire—the kind we explored in Chapter 3 and the ones you will learn about in this chapter and the next.

Basic Query Examples

In this section we'll walk through a few examples of queries on the Airline On-Time Performance and Causes of Flight Delays data set (*https://oreil.ly/gfzLZ*), which contains data on US flights including date, delay, distance, origin, and destination. It's available as a CSV file with over a million records. Using a schema, we'll read the data into a DataFrame and register the DataFrame as a temporary view (more on temporary views shortly) so we can query it with SQL.

Query examples are provided in code snippets, and Python and Scala notebooks containing all of the code presented here are available in the book's GitHub repo (*https://github.com/databricks/LearningSparkV2*). These examples will offer you a taste of how to use SQL in your Spark applications via the spark.sql programmatic interface (*https://spark.apache.org/sql*). Similar to the DataFrame API in its declarative flavor, this interface allows you to query structured data in your Spark applications.

Normally, in a standalone Spark application, you will create a SparkSession instance manually, as shown in the following example. However, in a Spark shell (or Databricks notebook), the SparkSession is created for you and accessible via the appropriately named variable spark.

Let's get started by reading the data set into a temporary view:

```scala
// In Scala
import org.apache.spark.sql.SparkSession
val spark = SparkSession
  .builder
  .appName("SparkSQLExampleApp")
  .getOrCreate()

// Path to data set
val csvFile="/databricks-datasets/learning-spark-v2/flights/departuredelays.csv"

// Read and create a temporary view
// Infer schema (note that for larger files you may want to specify the schema)
val df = spark.read.format("csv")
  .option("inferSchema", "true")
  .option("header", "true")
  .load(csvFile)
// Create a temporary view
df.createOrReplaceTempView("us_delay_flights_tbl")

# In Python
from pyspark.sql import SparkSession
# Create a SparkSession
spark = (SparkSession
  .builder
  .appName("SparkSQLExampleApp")
  .getOrCreate())
```

```
# Path to data set
csv_file = "/databricks-datasets/learning-spark-v2/flights/departuredelays.csv"

# Read and create a temporary view
# Infer schema (note that for larger files you
# may want to specify the schema)
df = (spark.read.format("csv")
  .option("inferSchema", "true")
  .option("header", "true")
  .load(csv_file))
df.createOrReplaceTempView("us_delay_flights_tbl")
```

If you want to specify a schema, you can use a DDL-formatted string. For example:

```
// In Scala
val schema = "date STRING, delay INT, distance INT,
 origin STRING, destination STRING"
```

```
# In Python
schema = "`date` STRING, `delay` INT, `distance` INT,
 `origin` STRING, `destination` STRING"
```

Now that we have a temporary view, we can issue SQL queries using Spark SQL. These queries are no different from those you might issue against a SQL table in, say, a MySQL or PostgreSQL database. The point here is to show that Spark SQL offers an ANSI:2003–compliant SQL interface, and to demonstrate the interoperability between SQL and DataFrames.

The US flight delays data set has five columns:

- The date column contains a string like 02190925. When converted, this maps to 02-19 09:25 am.
- The delay column gives the delay in minutes between the scheduled and actual departure times. Early departures show negative numbers.
- The distance column gives the distance in miles from the origin airport to the destination airport.
- The origin column contains the origin IATA airport code.
- The destination column contains the destination IATA airport code.

With that in mind, let's try some example queries against this data set.

First, we'll find all flights whose distance is greater than 1,000 miles:

```
spark.sql("""SELECT distance, origin, destination
FROM us_delay_flights_tbl WHERE distance > 1000
ORDER BY distance DESC""").show(10)
```

```
+--------+------+-----------+
|distance|origin|destination|
+--------+------+-----------+
|4330    |HNL   |JFK        |
|4330    |HNL   |JFK        |
|4330    |HNL   |JFK        |
|4330    |HNL   |JFK        |
|4330    |HNL   |JFK        |
|4330    |HNL   |JFK        |
|4330    |HNL   |JFK        |
|4330    |HNL   |JFK        |
|4330    |HNL   |JFK        |
|4330    |HNL   |JFK        |
+--------+------+-----------+
only showing top 10 rows
```

As the results show, all of the longest flights were between Honolulu (HNL) and New York (JFK). Next, we'll find all flights between San Francisco (SFO) and Chicago (ORD) with at least a two-hour delay:

```
spark.sql("""SELECT date, delay, origin, destination
FROM us_delay_flights_tbl
WHERE delay > 120 AND ORIGIN = 'SFO' AND DESTINATION = 'ORD'
ORDER by delay DESC""").show(10)
```

```
+--------+-----+------+-----------+
|date    |delay|origin|destination|
+--------+-----+------+-----------+
|02190925|1638 |SFO   |ORD        |
|01031755|396  |SFO   |ORD        |
|01022330|326  |SFO   |ORD        |
|01051205|320  |SFO   |ORD        |
|01190925|297  |SFO   |ORD        |
|02171115|296  |SFO   |ORD        |
|01071040|279  |SFO   |ORD        |
|01051550|274  |SFO   |ORD        |
|03120730|266  |SFO   |ORD        |
|01261104|258  |SFO   |ORD        |
+--------+-----+------+-----------+
only showing top 10 rows
```

It seems there were many significantly delayed flights between these two cities, on different dates. (As an exercise, convert the date column into a readable format and find the days or months when these delays were most common. Were the delays related to winter months or holidays?)

Let's try a more complicated query where we use the CASE clause in SQL. In the following example, we want to label all US flights, regardless of origin and destination, with an indication of the delays they experienced: Very Long Delays (> 6 hours),

Long Delays (2–6 hours), etc. We'll add these human-readable labels in a new column called `Flight_Delays`:

```
spark.sql("""SELECT delay, origin, destination,
              CASE
                  WHEN delay > 360 THEN 'Very Long Delays'
                  WHEN delay >= 120 AND delay <= 360 THEN 'Long Delays'
                  WHEN delay >= 60 AND delay < 120 THEN 'Short Delays'
                  WHEN delay > 0 and delay < 60 THEN 'Tolerable Delays'
                  WHEN delay = 0 THEN 'No Delays'
                  ELSE 'Early'
              END AS Flight_Delays
              FROM us_delay_flights_tbl
              ORDER BY origin, delay DESC""").show(10)
```

```
+-----+------+-----------+-------------+
|delay|origin|destination|Flight_Delays|
+-----+------+-----------+-------------+
|333  |ABE   |ATL        |Long Delays  |
|305  |ABE   |ATL        |Long Delays  |
|275  |ABE   |ATL        |Long Delays  |
|257  |ABE   |ATL        |Long Delays  |
|247  |ABE   |DTW        |Long Delays  |
|247  |ABE   |ATL        |Long Delays  |
|219  |ABE   |ORD        |Long Delays  |
|211  |ABE   |ATL        |Long Delays  |
|197  |ABE   |DTW        |Long Delays  |
|192  |ABE   |ORD        |Long Delays  |
+-----+------+-----------+-------------+
only showing top 10 rows
```

As with the DataFrame and Dataset APIs, with the `spark.sql` interface you can conduct common data analysis operations like those we explored in the previous chapter. The computations undergo an identical journey in the Spark SQL engine (see "The Catalyst Optimizer" on page 77 in Chapter 3 for details), giving you the same results.

All three of the preceding SQL queries can be expressed with an equivalent Data-Frame API query. For example, the first query can be expressed in the Python Data-Frame API as:

```
# In Python
from pyspark.sql.functions import col, desc
(df.select("distance", "origin", "destination")
  .where(col("distance") > 1000)
  .orderBy(desc("distance"))).show(10)

# Or
(df.select("distance", "origin", "destination")
  .where("distance > 1000")
  .orderBy("distance", ascending=False).show(10))
```

This produces the same results as the SQL query:

```
+--------+------+-----------+
|distance|origin|destination|
+--------+------+-----------+
|4330    |HNL   |JFK        |
|4330    |HNL   |JFK        |
|4330    |HNL   |JFK        |
|4330    |HNL   |JFK        |
|4330    |HNL   |JFK        |
|4330    |HNL   |JFK        |
|4330    |HNL   |JFK        |
|4330    |HNL   |JFK        |
|4330    |HNL   |JFK        |
|4330    |HNL   |JFK        |
+--------+------+-----------+
only showing top 10 rows
```

As an exercise, try converting the other two SQL queries to use the DataFrame API.

As these examples show, using the Spark SQL interface to query data is similar to writing a regular SQL query to a relational database table. Although the queries are in SQL, you can feel the similarity in readability and semantics to DataFrame API operations, which you encountered in Chapter 3 and will explore further in the next chapter.

To enable you to query structured data as shown in the preceding examples, Spark manages all the complexities of creating and managing views and tables, both in memory and on disk. That leads us to our next topic: how tables and views are created and managed.

SQL Tables and Views

Tables hold data. Associated with each table in Spark is its relevant metadata, which is information about the table and its data: the schema, description, table name, database name, column names, partitions, physical location where the actual data resides, etc. All of this is stored in a central metastore.

Instead of having a separate metastore for Spark tables, Spark by default uses the Apache Hive metastore, located at /user/hive/warehouse, to persist all the metadata about your tables. However, you may change the default location by setting the Spark config variable spark.sql.warehouse.dir to another location, which can be set to a local or external distributed storage.

Managed Versus UnmanagedTables

Spark allows you to create two types of tables: managed and unmanaged. For a *managed* table, Spark manages both the metadata and the data in the file store. This could be a local filesystem, HDFS, or an object store such as Amazon S3 or Azure Blob. For an *unmanaged* table, Spark only manages the metadata, while you manage the data yourself in an external data source (*https://oreil.ly/Scvor*) such as Cassandra.

With a managed table, because Spark manages everything, a SQL command such as `DROP TABLE table_name` deletes both the metadata and the data. With an unmanaged table, the same command will delete only the metadata, not the actual data. We will look at some examples of how to create managed and unmanaged tables in the next section.

Creating SQL Databases and Tables

Tables reside within a database. By default, Spark creates tables under the `default` database. To create your own database name, you can issue a SQL command from your Spark application or notebook. Using the US flight delays data set, let's create both a managed and an unmanaged table. To begin, we'll create a database called `learn_spark_db` and tell Spark we want to use that database:

```
// In Scala/Python
spark.sql("CREATE DATABASE learn_spark_db")
spark.sql("USE learn_spark_db")
```

From this point, any commands we issue in our application to create tables will result in the tables being created in this database and residing under the database name `learn_spark_db`.

Creating a managed table

To create a managed table within the database `learn_spark_db`, you can issue a SQL query like the following:

```
// In Scala/Python
spark.sql("CREATE TABLE managed_us_delay_flights_tbl (date STRING, delay INT,
    distance INT, origin STRING, destination STRING)")
```

You can do the same thing using the DataFrame API like this:

```
# In Python
# Path to our US flight delays CSV file
csv_file = "/databricks-datasets/learning-spark-v2/flights/departuredelays.csv"
# Schema as defined in the preceding example
schema="date STRING, delay INT, distance INT, origin STRING, destination STRING"
flights_df = spark.read.csv(csv_file, schema=schema)
flights_df.write.saveAsTable("managed_us_delay_flights_tbl")
```

Both of these statements will create the managed table `us_delay_flights_tbl` in the `learn_spark_db` database.

Creating an unmanaged table

By contrast, you can create unmanaged tables from your own data sources—say, Parquet, CSV, or JSON files stored in a file store accessible to your Spark application.

To create an unmanaged table from a data source such as a CSV file, in SQL use:

```
spark.sql("""CREATE TABLE us_delay_flights_tbl(date STRING, delay INT,
   distance INT, origin STRING, destination STRING)
   USING csv OPTIONS (PATH
   '/databricks-datasets/learning-spark-v2/flights/departuredelays.csv')""")
```

And within the DataFrame API use:

```
(flights_df
  .write
  .option("path", "/tmp/data/us_flights_delay")
  .saveAsTable("us_delay_flights_tbl"))
```

 To enable you to explore these examples, we have created Python and Scala example notebooks that you can find in the book's GitHub repo (*https://github.com/databricks/LearningSparkV2*).

Creating Views

In addition to creating tables, Spark can create views on top of existing tables. Views can be global (visible across all SparkSessions on a given cluster) or session-scoped (visible only to a single SparkSession), and they are temporary: they disappear after your Spark application terminates.

Creating views (*https://oreil.ly/8OqlM*) has a similar syntax to creating tables within a database. Once you create a view, you can query it as you would a table. The difference between a view and a table is that views don't actually hold the data; tables persist after your Spark application terminates, but views disappear.

You can create a view from an existing table using SQL. For example, if you wish to work on only the subset of the US flight delays data set with origin airports of New York (JFK) and San Francisco (SFO), the following queries will create global temporary and temporary views consisting of just that slice of the table:

```
-- In SQL
CREATE OR REPLACE GLOBAL TEMP VIEW us_origin_airport_SFO_global_tmp_view AS
   SELECT date, delay, origin, destination from us_delay_flights_tbl WHERE
   origin = 'SFO';

CREATE OR REPLACE TEMP VIEW us_origin_airport_JFK_tmp_view AS
   SELECT date, delay, origin, destination from us_delay_flights_tbl WHERE
   origin = 'JFK'
```

You can accomplish the same thing with the DataFrame API as follows:

```python
# In Python
df_sfo = spark.sql("SELECT date, delay, origin, destination FROM
  us_delay_flights_tbl WHERE origin = 'SFO'")
df_jfk = spark.sql("SELECT date, delay, origin, destination FROM
  us_delay_flights_tbl WHERE origin = 'JFK'")

# Create a temporary and global temporary view
df_sfo.createOrReplaceGlobalTempView("us_origin_airport_SFO_global_tmp_view")
df_jfk.createOrReplaceTempView("us_origin_airport_JFK_tmp_view")
```

Once you've created these views, you can issue queries against them just as you would against a table. Keep in mind that when accessing a global temporary view you must use the prefix global_temp.<view_name>, because Spark creates global temporary views in a global temporary database called global_temp. For example:

```sql
-- In SQL
SELECT * FROM global_temp.us_origin_airport_SFO_global_tmp_view
```

By contrast, you can access the normal temporary view without the global_temp prefix:

```sql
-- In SQL
SELECT * FROM us_origin_airport_JFK_tmp_view
```

```scala
// In Scala/Python
spark.read.table("us_origin_airport_JFK_tmp_view")
// Or
spark.sql("SELECT * FROM us_origin_airport_JFK_tmp_view")
```

You can also drop a view just like you would a table:

```sql
-- In SQL
DROP VIEW IF EXISTS us_origin_airport_SFO_global_tmp_view;
DROP VIEW IF EXISTS us_origin_airport_JFK_tmp_view
```

```scala
// In Scala/Python
spark.catalog.dropGlobalTempView("us_origin_airport_SFO_global_tmp_view")
spark.catalog.dropTempView("us_origin_airport_JFK_tmp_view")
```

Temporary views versus global temporary views

The difference between *temporary* and *global temporary* views being subtle, it can be a source of mild confusion among developers new to Spark. A temporary view is tied to a single SparkSession within a Spark application. In contrast, a global temporary view is visible across multiple SparkSessions within a Spark application. Yes, you can create multiple SparkSessions (*https://oreil.ly/YbTFa*) within a single Spark application—this can be handy, for example, in cases where you want to access (and combine) data from two different SparkSessions that don't share the same Hive metastore configurations.

Viewing the Metadata

As mentioned previously, Spark manages the metadata associated with each managed or unmanaged table. This is captured in the `Catalog` (*https://oreil.ly/56HYV*), a high-level abstraction in Spark SQL for storing metadata. The `Catalog`'s functionality was expanded in Spark 2.x with new public methods enabling you to examine the metadata associated with your databases, tables, and views. Spark 3.0 extends it to use external `catalog` (which we briefly discuss in Chapter 12).

For example, within a Spark application, after creating the `SparkSession` variable `spark`, you can access all the stored metadata through methods like these:

```
// In Scala/Python
spark.catalog.listDatabases()
spark.catalog.listTables()
spark.catalog.listColumns("us_delay_flights_tbl")
```

Import the notebook from the book's GitHub repo (*https://github.com/databricks/Learn ingSparkV2*) and give it a try.

Caching SQL Tables

Although we will discuss table caching strategies in the next chapter, it's worth mentioning here that, like DataFrames, you can cache and uncache SQL tables and views. In Spark 3.0 (*https://oreil.ly/2ptwu*), in addition to other options, you can specify a table as LAZY, meaning that it should only be cached when it is first used instead of immediately:

```
-- In SQL
CACHE [LAZY] TABLE <table-name>
UNCACHE TABLE <table-name>
```

Reading Tables into DataFrames

Often, data engineers build data pipelines as part of their regular data ingestion and ETL processes. They populate Spark SQL databases and tables with cleansed data for consumption by applications downstream.

Let's assume you have an existing database, `learn_spark_db`, and table, `us_delay_flights_tbl`, ready for use. Instead of reading from an external JSON file, you can simply use SQL to query the table and assign the returned result to a DataFrame:

```
// In Scala
val usFlightsDF = spark.sql("SELECT * FROM us_delay_flights_tbl")
val usFlightsDF2 = spark.table("us_delay_flights_tbl")
```

```
# In Python
us_flights_df = spark.sql("SELECT * FROM us_delay_flights_tbl")
us_flights_df2 = spark.table("us_delay_flights_tbl")
```

Now you have a cleansed DataFrame read from an existing Spark SQL table. You can also read data in other formats using Spark's built-in data sources, giving you the flexibility to interact with various common file formats.

Data Sources for DataFrames and SQL Tables

As shown in Figure 4-1, Spark SQL provides an interface to a variety of data sources. It also provides a set of common methods for reading and writing data to and from these data sources using the Data Sources API (*https://oreil.ly/_8-6A*).

In this section we will cover some of the built-in data sources (*https://oreil.ly/Hj9pd*), available file formats, and ways to load and write data, along with specific options pertaining to these data sources. But first, let's take a closer look at two high-level Data Source API constructs that dictate the manner in which you interact with different data sources: DataFrameReader and DataFrameWriter.

DataFrameReader

DataFrameReader (*https://oreil.ly/UZXdx*) is the core construct for reading data from a data source into a DataFrame. It has a defined format and a recommended pattern for usage:

```
DataFrameReader.format(args).option("key", "value").schema(args).load()
```

This pattern of stringing methods together is common in Spark, and easy to read. We saw it in Chapter 3 when exploring common data analysis patterns.

Note that you can only access a DataFrameReader through a SparkSession instance. That is, you cannot create an instance of DataFrameReader. To get an instance handle to it, use:

```
SparkSession.read
// or
SparkSession.readStream
```

While read returns a handle to DataFrameReader to read into a DataFrame from a static data source, readStream returns an instance to read from a streaming source. (We will cover Structured Streaming later in the book.)

Arguments to each of the public methods to DataFrameReader take different values. Table 4-1 enumerates these, with a subset of the supported arguments.

Table 4-1. DataFrameReader methods, arguments, and options

Method	Arguments	Description
format()	"parquet","csv","txt","json", "jdbc", "orc", "avro", etc.	If you don't specify this method, then the default is Parquet or whatever is set in spark.sql.sour ces.default.
option()	("mode", {PERMISSIVE | FAILFAST | DROPMALFORMED }) ("inferSchema", {true | false}) ("path", "path_file_data_source")	A series of key/value pairs and options. The Spark documentation (*https://oreil.ly/XujEK*) shows some examples and explains the different modes and their actions. The default mode is PERMISSIVE. The "infer Schema" and "mode" options are specific to the JSON and CSV file formats.
schema()	DDL String or StructType, e.g., 'A INT, B STRING' or StructType(...)	For JSON or CSV format, you can specify to infer the schema in the option() method. Generally, providing a schema for any format makes loading faster and ensures your data conforms to the expected schema.
load()	"/path/to/data/source"	The path to the data source. This can be empty if specified in option("path", "...").

While we won't comprehensively enumerate all the different combinations of arguments and options, the documentation for Python, Scala, R, and Java (*https://oreil.ly/RsfRg*) offers suggestions and guidance. It's worthwhile to show a couple of examples, though:

```scala
// In Scala
// Use Parquet
val file = """/databricks-datasets/learning-spark-v2/flights/summary-
  data/parquet/2010-summary.parquet"""
val df = spark.read.format("parquet").load(file)
// Use Parquet; you can omit format("parquet") if you wish as it's the default
val df2 = spark.read.load(file)
// Use CSV
val df3 = spark.read.format("csv")
  .option("inferSchema", "true")
  .option("header", "true")
  .option("mode", "PERMISSIVE")
  .load("/databricks-datasets/learning-spark-v2/flights/summary-data/csv/*")
// Use JSON
val df4 = spark.read.format("json")
  .load("/databricks-datasets/learning-spark-v2/flights/summary-data/json/*")
```

In general, no schema is needed when reading from a static Parquet data source—the Parquet metadata usually contains the schema, so it's inferred. However, for streaming data sources you will have to provide a schema. (We will cover reading from streaming data sources in Chapter 8.)

Parquet is the default and preferred data source for Spark because it's efficient, uses columnar storage, and employs a fast compression algorithm. You will see additional benefits later (such as columnar pushdown), when we cover the Catalyst optimizer in greater depth.

DataFrameWriter

DataFrameWriter (*https://oreil.ly/SM1LR*) does the reverse of its counterpart: it saves or writes data to a specified built-in data source. Unlike with `DataFrameReader`, you access its instance not from a `SparkSession` but from the DataFrame you wish to save. It has a few recommended usage patterns:

```
DataFrameWriter.format(args)
  .option(args)
  .bucketBy(args)
  .partitionBy(args)
  .save(path)
```

```
DataFrameWriter.format(args).option(args).sortBy(args).saveAsTable(table)
```

To get an instance handle, use:

```
DataFrame.write
// or
DataFrame.writeStream
```

Arguments to each of the methods to `DataFrameWriter` also take different values. We list these in Table 4-2, with a subset of the supported arguments.

Table 4-2. DataFrameWriter methods, arguments, and options

Method	Arguments	Description
`format()`	`"parquet"`, `"csv"`, `"txt"`, `"json"`, `"jdbc"`, `"orc"`, `"avro"`, etc.	If you don't specify this method, then the default is Parquet or whatever is set in `spark.sql.sources.default`.
`option()`	`("mode", {append | overwrite | ignore | error or errorifex ists} )` `("mode", {SaveMode.Overwrite | SaveMode.Append, Save Mode.Ignore, SaveMode.ErrorI fExists})` `("path", "path_to_write_to")`	A series of key/value pairs and options. The Spark documentation (*https://oreil.ly/w7J0l*) shows some examples. This is an overloaded method. The default mode options are `error or errorifexists` and `SaveMode.ErrorI fExists`; they throw an exception at runtime if the data already exists.

Method	Arguments	Description
buck etBy()	(numBuckets, col, col..., coln)	The number of buckets and names of columns to bucket by. Uses Hive's bucketing scheme on a filesystem.
save()	"/path/to/data/source"	The path to save to. This can be empty if specified in option("path", "...").
saveAsTa ble()	"table_name"	The table to save to.

Here's a short example snippet to illustrate the use of methods and arguments:

```scala
// In Scala
// Use JSON
val location = ...
df.write.format("json").mode("overwrite").save(location)
```

Parquet

We'll start our exploration of data sources with Parquet (*https://oreil.ly/-wptz*), because it's the default data source in Spark. Supported and widely used by many big data processing frameworks and platforms, Parquet is an open source columnar file format that offers many I/O optimizations (such as compression, which saves storage space and allows for quick access to data columns).

Because of its efficiency and these optimizations, we recommend that after you have transformed and cleansed your data, you save your DataFrames in the Parquet format for downstream consumption. (Parquet is also the default table open format for Delta Lake, which we will cover in Chapter 9.)

Reading Parquet files into a DataFrame

Parquet files (*https://oreil.ly/CTVzK*) are stored in a directory structure that contains the data files, metadata, a number of compressed files, and some status files. Metadata in the footer contains the version of the file format, the schema, and column data such as the path, etc.

For example, a directory in a Parquet file might contain a set of files like this:

```
_SUCCESS
_committed_1799640464332036264
_started_1799640464332036264
part-00000-tid-1799640464332036264-91273258-d7ef-4dc7-<...>-c000.snappy.parquet
```

There may be a number of *part-XXXX* compressed files in a directory (the names shown here have been shortened to fit on the page).

To read Parquet files into a DataFrame, you simply specify the format and path:

```
// In Scala
val file = """/databricks-datasets/learning-spark-v2/flights/summary-data/
  parquet/2010-summary.parquet/"""
val df = spark.read.format("parquet").load(file)

# In Python
file = """/databricks-datasets/learning-spark-v2/flights/summary-data/parquet/
  2010-summary.parquet/"""
df = spark.read.format("parquet").load(file)
```

Unless you are reading from a streaming data source there's no need to supply the schema, because Parquet saves it as part of its metadata.

Reading Parquet files into a Spark SQL table

As well as reading Parquet files into a Spark DataFrame, you can also create a Spark SQL unmanaged table or view directly using SQL:

```
-- In SQL
CREATE OR REPLACE TEMPORARY VIEW us_delay_flights_tbl
    USING parquet
    OPTIONS (
      path "/databricks-datasets/learning-spark-v2/flights/summary-data/parquet/
      2010-summary.parquet/" )
```

Once you've created the table or view, you can read data into a DataFrame using SQL, as we saw in some earlier examples:

```
// In Scala
spark.sql("SELECT * FROM us_delay_flights_tbl").show()

# In Python
spark.sql("SELECT * FROM us_delay_flights_tbl").show()
```

Both of these operations return the same results:

```
+-----------------+-------------------+-----+
|DEST_COUNTRY_NAME|ORIGIN_COUNTRY_NAME|count|
+-----------------+-------------------+-----+
|United States    |Romania            |1    |
|United States    |Ireland            |264  |
|United States    |India              |69   |
|Egypt            |United States      |24   |
|Equatorial Guinea|United States      |1    |
|United States    |Singapore          |25   |
|United States    |Grenada            |54   |
|Costa Rica       |United States      |477  |
|Senegal          |United States      |29   |
|United States    |Marshall Islands   |44   |
+-----------------+-------------------+-----+
only showing top 10 rows
```

Writing DataFrames to Parquet files

Writing or saving a DataFrame as a table or file is a common operation in Spark. To write a DataFrame you simply use the methods and arguments to the DataFrame Writer outlined earlier in this chapter, supplying the location to save the Parquet files to. For example:

```
// In Scala
df.write.format("parquet")
  .mode("overwrite")
  .option("compression", "snappy")
  .save("/tmp/data/parquet/df_parquet")
```

```
# In Python
(df.write.format("parquet")
  .mode("overwrite")
  .option("compression", "snappy")
  .save("/tmp/data/parquet/df_parquet"))
```

> Recall that Parquet is the default file format. If you don't include the format() method, the DataFrame will still be saved as a Parquet file.

This will create a set of compact and compressed Parquet files at the specified path. Since we used snappy as our compression choice here, we'll have snappy compressed files. For brevity, this example generated only one file; normally, there may be a dozen or so files created:

```
-rw-r--r--  1 jules  wheel    0 May 19 10:58 _SUCCESS
-rw-r--r--  1 jules  wheel  966 May 19 10:58 part-00000-<...>-c000.snappy.parquet
```

Writing DataFrames to Spark SQL tables

Writing a DataFrame to a SQL table is as easy as writing to a file—just use saveAsTable() instead of save(). This will create a managed table called us_delay_flights_tbl:

```
// In Scala
df.write
  .mode("overwrite")
  .saveAsTable("us_delay_flights_tbl")
```

```
# In Python
(df.write
  .mode("overwrite")
  .saveAsTable("us_delay_flights_tbl"))
```

To sum up, Parquet is the preferred and default built-in data source file format in Spark, and it has been adopted by many other frameworks. We recommend that you use this format in your ETL and data ingestion processes.

JSON

JavaScript Object Notation (JSON) is also a popular data format. It came to prominence as an easy-to-read and easy-to-parse format compared to XML. It has two representational formats: single-line mode and multiline mode (*https://oreil.ly/bBdLc*). Both modes are supported in Spark.

In single-line mode each line denotes a single JSON object (*http://jsonlines.org/*), whereas in multiline mode the entire multiline object constitutes a single JSON object. To read in this mode, set multiLine to true in the option() method.

Reading a JSON file into a DataFrame

You can read a JSON file into a DataFrame the same way you did with Parquet—just specify "json" in the format() method:

```scala
// In Scala
val file = "/databricks-datasets/learning-spark-v2/flights/summary-data/json/*"
val df = spark.read.format("json").load(file)
```

```python
# In Python
file = "/databricks-datasets/learning-spark-v2/flights/summary-data/json/*"
df = spark.read.format("json").load(file)
```

Reading a JSON file into a Spark SQL table

You can also create a SQL table from a JSON file just like you did with Parquet:

```sql
-- In SQL
CREATE OR REPLACE TEMPORARY VIEW us_delay_flights_tbl
    USING json
    OPTIONS (
      path "/databricks-datasets/learning-spark-v2/flights/summary-data/json/*"
    )
```

Once the table is created, you can read data into a DataFrame using SQL:

```
// In Scala/Python
spark.sql("SELECT * FROM us_delay_flights_tbl").show()

+-----------------+-------------------+-----+
|DEST_COUNTRY_NAME|ORIGIN_COUNTRY_NAME|count|
+-----------------+-------------------+-----+
|United States    |Romania            |15   |
|United States    |Croatia            |1    |
|United States    |Ireland            |344  |
|Egypt            |United States      |15   |
```

```
|United States  |India          |62   |
|United States  |Singapore      |1    |
|United States  |Grenada        |62   |
|Costa Rica     |United States  |588  |
|Senegal        |United States  |40   |
|Moldova        |United States  |1    |
+---------------+---------------+-----+
only showing top 10 rows
```

Writing DataFrames to JSON files

Saving a DataFrame as a JSON file is simple. Specify the appropriate
DataFrameWriter methods and arguments, and supply the location to save the JSON
files to:

```scala
// In Scala
df.write.format("json")
  .mode("overwrite")
  .option("compression", "snappy")
  .save("/tmp/data/json/df_json")
```

```python
# In Python
(df.write.format("json")
  .mode("overwrite")
  .option("compression", "snappy")
  .save("/tmp/data/json/df_json"))
```

This creates a directory at the specified path populated with a set of compact JSON
files:

```
-rw-r--r--  1 jules  wheel   0 May 16 14:44 _SUCCESS
-rw-r--r--  1 jules  wheel  71 May 16 14:44 part-00000-<...>-c000.json
```

JSON data source options

Table 4-3 describes common JSON options for DataFrameReader (*https://oreil.ly/
iDZ2T*) and DataFrameWriter (*https://oreil.ly/MunK1*). For a comprehensive list, we
refer you to the documentation.

Table 4-3. JSON options for DataFrameReader and DataFrameWriter

Property name	Values	Meaning	Scope
compression	none, uncompressed, bzip2, deflate, gzip, lz4, or snappy	Use this compression codec for writing. Note that read will only detect the compression or codec from the file extension.	Write
dateFormat	yyyy-MM-dd or DateTimeFormatter	Use this format or any format from Java's DateTime Formatter.	Read/ write
multiLine	true, false	Use multiline mode. Default is false (single-line mode).	Read

Property name	Values	Meaning	Scope
allowUnquoted FieldNames	true, false	Allow unquoted JSON field names. Default is false.	Read

CSV

As widely used as plain text files, this common text file format captures each datum or field delimited by a comma; each line with comma-separated fields represents a record. Even though a comma is the default separator, you may use other delimiters to separate fields in cases where commas are part of your data. Popular spreadsheets can generate CSV files, so it's a popular format among data and business analysts.

Reading a CSV file into a DataFrame

As with the other built-in data sources, you can use the DataFrameReader methods and arguments to read a CSV file into a DataFrame:

```scala
// In Scala
val file = "/databricks-datasets/learning-spark-v2/flights/summary-data/csv/*"
val schema = "DEST_COUNTRY_NAME STRING, ORIGIN_COUNTRY_NAME STRING, count INT"

val df = spark.read.format("csv")
  .schema(schema)
  .option("header", "true")
  .option("mode", "FAILFAST")    // Exit if any errors
  .option("nullValue", "")       // Replace any null data with quotes
  .load(file)
```

```python
# In Python
file = "/databricks-datasets/learning-spark-v2/flights/summary-data/csv/*"
schema = "DEST_COUNTRY_NAME STRING, ORIGIN_COUNTRY_NAME STRING, count INT"
df = (spark.read.format("csv")
  .option("header", "true")
  .schema(schema)
  .option("mode", "FAILFAST")  # Exit if any errors
  .option("nullValue", "")       # Replace any null data field with quotes
  .load(file))
```

Reading a CSV file into a Spark SQL table

Creating a SQL table from a CSV data source is no different from using Parquet or JSON:

```sql
-- In SQL
CREATE OR REPLACE TEMPORARY VIEW us_delay_flights_tbl
    USING csv
    OPTIONS (
      path "/databricks-datasets/learning-spark-v2/flights/summary-data/csv/*",
      header "true",
      inferSchema "true",
```

```
    mode "FAILFAST"
)
```

Once you've created the table, you can read data into a DataFrame using SQL as
before:

```
// In Scala/Python
spark.sql("SELECT * FROM us_delay_flights_tbl").show(10)
```

```
+------------------+-------------------+-----+
|DEST_COUNTRY_NAME |ORIGIN_COUNTRY_NAME|count|
+------------------+-------------------+-----+
|United States     |Romania            |1    |
|United States     |Ireland            |264  |
|United States     |India              |69   |
|Egypt             |United States      |24   |
|Equatorial Guinea |United States      |1    |
|United States     |Singapore          |25   |
|United States     |Grenada            |54   |
|Costa Rica        |United States      |477  |
|Senegal           |United States      |29   |
|United States     |Marshall Islands   |44   |
+------------------+-------------------+-----+
only showing top 10 rows
```

Writing DataFrames to CSV files

Saving a DataFrame as a CSV file is simple. Specify the appropriate DataFrameWriter
methods and arguments, and supply the location to save the CSV files to:

```
// In Scala
df.write.format("csv").mode("overwrite").save("/tmp/data/csv/df_csv")
```

```
# In Python
df.write.format("csv").mode("overwrite").save("/tmp/data/csv/df_csv")
```

This generates a folder at the specified location, populated with a bunch of com‐
pressed and compact files:

```
-rw-r--r--  1 jules  wheel   0 May 16 12:17 _SUCCESS
-rw-r--r--  1 jules  wheel  36 May 16 12:17 part-00000-251690eb-<...>-c000.csv
```

CSV data source options

Table 4-4 describes some of the common CSV options for DataFrameReader (*https://
oreil.ly/Au6Kd*) and DataFrameWriter (*https://oreil.ly/4g-vz*). Because CSV files can
be complex, many options are available; for a comprehensive list we refer you to the
documentation.

Table 4-4. CSV options for DataFrameReader and DataFrameWriter

Property name	Values	Meaning	Scope
`compression`	`none, bzip2, deflate,` `gzip, lz4,` or `snappy`	Use this compression codec for writing.	Write
`dateFormat`	`yyyy-MM-dd` or `DateTime` `Formatter`	Use this format or any format from Java's `Date` `TimeFormatter`.	Read/ write
`multiLine`	`true, false`	Use multiline mode. Default is `false` (single-line mode).	Read
`inferSchema`	`true, false`	If `true`, Spark will determine the column data types. Default is `false`.	Read
`sep`	Any character	Use this character to separate column values in a row. Default delimiter is a comma (`,`).	Read/ write
`escape`	Any character	Use this character to escape quotes. Default is `\`.	Read/ write
`header`	`true, false`	Indicates whether the first line is a header denoting each column name. Default is `false`.	Read/ write

Avro

Introduced in Spark 2.4 (*https://oreil.ly/gqZl0*) as a built-in data source, the Avro format (*https://oreil.ly/UaJoR*) is used, for example, by Apache Kafka (*https://oreil.ly/jhdTI*) for message serializing and deserializing. It offers many benefits, including direct mapping to JSON, speed and efficiency, and bindings available for many programming languages.

Reading an Avro file into a DataFrame

Reading an Avro file into a DataFrame using `DataFrameReader` is consistent in usage with the other data sources we have discussed in this section:

```scala
// In Scala
val df = spark.read.format("avro")
 .load("/databricks-datasets/learning-spark-v2/flights/summary-data/avro/*")
df.show(false)
```

```python
# In Python
df = (spark.read.format("avro")
 .load("/databricks-datasets/learning-spark-v2/flights/summary-data/avro/*"))
df.show(truncate=False)
```

```
+-----------------+-------------------+-----+
|DEST_COUNTRY_NAME|ORIGIN_COUNTRY_NAME|count|
+-----------------+-------------------+-----+
|United States    |Romania            |1    |
|United States    |Ireland            |264  |
|United States    |India              |69   |
|Egypt            |United States      |24   |
```

```
|Equatorial Guinea|United States       |1   |
|United States    |Singapore           |25  |
|United States    |Grenada             |54  |
|Costa Rica       |United States       |477 |
|Senegal          |United States       |29  |
|United States    |Marshall Islands    |44  |
+-----------------+--------------------+-----+
only showing top 10 rows
```

Reading an Avro file into a Spark SQL table

Again, creating SQL tables using an Avro data source is no different from using Parquet, JSON, or CSV:

```
-- In SQL
CREATE OR REPLACE TEMPORARY VIEW episode_tbl
    USING avro
    OPTIONS (
      path "/databricks-datasets/learning-spark-v2/flights/summary-data/avro/*"
    )
```

Once you've created a table, you can read data into a DataFrame using SQL:

```
// In Scala
spark.sql("SELECT * FROM episode_tbl").show(false)
```

```
# In Python
spark.sql("SELECT * FROM episode_tbl").show(truncate=False)
```

```
+-----------------+--------------------+-----+
|DEST_COUNTRY_NAME|ORIGIN_COUNTRY_NAME |count|
+-----------------+--------------------+-----+
|United States    |Romania             |1   |
|United States    |Ireland             |264 |
|United States    |India               |69  |
|Egypt            |United States       |24  |
|Equatorial Guinea|United States       |1   |
|United States    |Singapore           |25  |
|United States    |Grenada             |54  |
|Costa Rica       |United States       |477 |
|Senegal          |United States       |29  |
|United States    |Marshall Islands    |44  |
+-----------------+--------------------+-----+
only showing top 10 rows
```

Writing DataFrames to Avro files

Writing a DataFrame as an Avro file is simple. As usual, specify the appropriate Data
FrameWriter methods and arguments, and supply the location to save the Avro files
to:

```
// In Scala
df.write
  .format("avro")
  .mode("overwrite")
  .save("/tmp/data/avro/df_avro")

# In Python
(df.write
  .format("avro")
  .mode("overwrite")
  .save("/tmp/data/avro/df_avro"))
```

This generates a folder at the specified location, populated with a bunch of compressed and compact files:

```
-rw-r--r--  1 jules  wheel    0 May 17 11:54 _SUCCESS
-rw-r--r--  1 jules  wheel  526 May 17 11:54 part-00000-ffdf70f4-<...>-c000.avro
```

Avro data source options

Table 4-5 describes common options for `DataFrameReader` and `DataFrameWriter`. A comprehensive list of options is in the documentation (*https://oreil.ly/Jvrd_*).

Table 4-5. Avro options for DataFrameReader and DataFrameWriter

Property name	Default value	Meaning	Scope
avroSchema	None	Optional Avro schema provided by a user in JSON format. The data type and naming of record fields should match the input Avro data or Catalyst data (Spark internal data type), otherwise the read/write action will fail.	Read/write
recordName	topLevel Record	Top-level record name in write result, which is required in the Avro spec.	Write
recordNamespace	""	Record namespace in write result.	Write
ignoreExtension	true	If this option is enabled, all files (with and without the *.avro* extension) are loaded. Otherwise, files without the *.avro* extension are ignored.	Read
compression	snappy	Allows you to specify the compression codec to use in writing. Currently supported codecs are uncompressed, snappy, deflate, bzip2, and xz. If this option is not set, the value in spark.sql.avro.compression.codec is taken into account.	Write

ORC

As an additional optimized columnar file format, Spark 2.x supports a vectorized ORC reader (*https://oreil.ly/N_Brd*). Two Spark configurations dictate which ORC implementation to use. When `spark.sql.orc.impl` is set to `native` and `spark.sql.orc.enableVectorizedReader` is set to `true`, Spark uses the vectorized

ORC reader. A vectorized reader (*https://oreil.ly/E2xiZ*) reads blocks of rows (often 1,024 per block) instead of one row at a time, streamlining operations and reducing CPU usage for intensive operations like scans, filters, aggregations, and joins.

For Hive ORC SerDe (serialization and deserialization) tables created with the SQL command `USING HIVE OPTIONS (fileFormat 'ORC')`, the vectorized reader is used when the Spark configuration parameter `spark.sql.hive.convertMetastoreOrc` is set to `true`.

Reading an ORC file into a DataFrame

To read in a DataFrame using the ORC vectorized reader, you can just use the normal `DataFrameReader` methods and options:

```scala
// In Scala
val file = "/databricks-datasets/learning-spark-v2/flights/summary-data/orc/*"
val df = spark.read.format("orc").load(file)
df.show(10, false)
```

```python
# In Python
file = "/databricks-datasets/learning-spark-v2/flights/summary-data/orc/*"
df = spark.read.format("orc").option("path", file).load()
df.show(10, False)
```

```
+-----------------+-------------------+-----+
|DEST_COUNTRY_NAME|ORIGIN_COUNTRY_NAME|count|
+-----------------+-------------------+-----+
|United States    |Romania            |1    |
|United States    |Ireland            |264  |
|United States    |India              |69   |
|Egypt            |United States      |24   |
|Equatorial Guinea|United States      |1    |
|United States    |Singapore          |25   |
|United States    |Grenada            |54   |
|Costa Rica       |United States      |477  |
|Senegal          |United States      |29   |
|United States    |Marshall Islands   |44   |
+-----------------+-------------------+-----+
only showing top 10 rows
```

Reading an ORC file into a Spark SQL table

There is no difference from Parquet, JSON, CSV, or Avro when creating a SQL view using an ORC data source:

```sql
-- In SQL
CREATE OR REPLACE TEMPORARY VIEW us_delay_flights_tbl
    USING orc
    OPTIONS (
      path "/databricks-datasets/learning-spark-v2/flights/summary-data/orc/*"
    )
```

Once a table is created, you can read data into a DataFrame using SQL as usual:

```
// In Scala/Python
spark.sql("SELECT * FROM us_delay_flights_tbl").show()
```

```
+-----------------+-------------------+-----+
|DEST_COUNTRY_NAME|ORIGIN_COUNTRY_NAME|count|
+-----------------+-------------------+-----+
|United States    |Romania            |1    |
|United States    |Ireland            |264  |
|United States    |India              |69   |
|Egypt            |United States      |24   |
|Equatorial Guinea|United States      |1    |
|United States    |Singapore          |25   |
|United States    |Grenada            |54   |
|Costa Rica       |United States      |477  |
|Senegal          |United States      |29   |
|United States    |Marshall Islands   |44   |
+-----------------+-------------------+-----+
only showing top 10 rows
```

Writing DataFrames to ORC files

Writing back a transformed DataFrame after reading is equally simple using the `DataFrameWriter` methods:

```
// In Scala
df.write.format("orc")
  .mode("overwrite")
  .option("compression", "snappy")
  .save("/tmp/data/orc/df_orc")
```

```
# In Python
(df.write.format("orc")
  .mode("overwrite")
  .option("compression", "snappy")
  .save("/tmp/data/orc/flights_orc"))
```

The result will be a folder at the specified location containing some compressed ORC files:

```
-rw-r--r-- 1 jules  wheel    0 May 16 17:23 _SUCCESS
-rw-r--r-- 1 jules  wheel  547 May 16 17:23 part-00000-<...>-c000.snappy.orc
```

Images

In Spark 2.4 the community introduced a new data source, image files (*https://oreil.ly/ JfKBD*), to support deep learning and machine learning frameworks such as Tensor-Flow and PyTorch. For computer vision–based machine learning applications, loading and processing image data sets is important.

Reading an image file into a DataFrame

As with all of the previous file formats, you can use the `DataFrameReader` methods and options to read in an image file as shown here:

```scala
// In Scala
import org.apache.spark.ml.source.image

val imageDir = "/databricks-datasets/learning-spark-v2/cctvVideos/train_images/"
val imagesDF = spark.read.format("image").load(imageDir)

imagesDF.printSchema

imagesDF.select("image.height", "image.width", "image.nChannels", "image.mode",
  "label").show(5, false)
```

```python
# In Python
from pyspark.ml import image

image_dir = "/databricks-datasets/learning-spark-v2/cctvVideos/train_images/"
images_df = spark.read.format("image").load(image_dir)
images_df.printSchema()
```

```
root
 |-- image: struct (nullable = true)
 |    |-- origin: string (nullable = true)
 |    |-- height: integer (nullable = true)
 |    |-- width: integer (nullable = true)
 |    |-- nChannels: integer (nullable = true)
 |    |-- mode: integer (nullable = true)
 |    |-- data: binary (nullable = true)
 |-- label: integer (nullable = true)
```

```python
images_df.select("image.height", "image.width", "image.nChannels", "image.mode",
  "label").show(5, truncate=False)
```

```
+------+-----+---------+----+-----+
|height|width|nChannels|mode|label|
+------+-----+---------+----+-----+
|288   |384  |3        |16  |0    |
|288   |384  |3        |16  |1    |
|288   |384  |3        |16  |0    |
|288   |384  |3        |16  |0    |
|288   |384  |3        |16  |0    |
+------+-----+---------+----+-----+
only showing top 5 rows
```

Binary Files

Spark 3.0 adds support for binary files as a data source (*https://oreil.ly/UXHZl*). The `DataFrameReader` converts each binary file into a single DataFrame row (record) that contains the raw content and metadata of the file. The binary file data source produces a DataFrame with the following columns:

- path: StringType
- modificationTime: TimestampType
- length: LongType
- content: BinaryType

Reading a binary file into a DataFrame

To read binary files, specify the data source format as a `binaryFile`. You can load files with paths matching a given global pattern while preserving the behavior of partition discovery with the data source option `pathGlobFilter`. For example, the following code reads all JPG files from the input directory with any partitioned directories:

```scala
// In Scala
val path = "/databricks-datasets/learning-spark-v2/cctvVideos/train_images/"
val binaryFilesDF = spark.read.format("binaryFile")
  .option("pathGlobFilter", "*.jpg")
  .load(path)
binaryFilesDF.show(5)
```

```python
# In Python
path = "/databricks-datasets/learning-spark-v2/cctvVideos/train_images/"
binary_files_df = (spark.read.format("binaryFile")
  .option("pathGlobFilter", "*.jpg")
  .load(path))
binary_files_df.show(5)
```

```
+--------------------+--------------------+------+--------------------+-----+
|                path|    modificationTime|length|             content|label|
+--------------------+--------------------+------+--------------------+-----+
|file:/Users/jules...|2020-02-12 12:04:24| 55037|[FF D8 FF E0 00 1...|    0|
|file:/Users/jules...|2020-02-12 12:04:24| 54634|[FF D8 FF E0 00 1...|    1|
|file:/Users/jules...|2020-02-12 12:04:24| 54624|[FF D8 FF E0 00 1...|    0|
|file:/Users/jules...|2020-02-12 12:04:24| 54505|[FF D8 FF E0 00 1...|    0|
|file:/Users/jules...|2020-02-12 12:04:24| 54475|[FF D8 FF E0 00 1...|    0|
+--------------------+--------------------+------+--------------------+-----+
only showing top 5 rows
```

To ignore partitioning data discovery in a directory, you can set `recursiveFileLookup` to `"true"`:

```scala
// In Scala
val binaryFilesDF = spark.read.format("binaryFile")
  .option("pathGlobFilter", "*.jpg")
  .option("recursiveFileLookup", "true")
  .load(path)
binaryFilesDF.show(5)
```

```python
# In Python
binary_files_df = (spark.read.format("binaryFile")
  .option("pathGlobFilter", "*.jpg")
  .option("recursiveFileLookup", "true")
  .load(path))
binary_files_df.show(5)
```

```
+--------------------+-------------------+------+--------------------+
|                path|   modificationTime|length|             content|
+--------------------+-------------------+------+--------------------+
|file:/Users/jules...|2020-02-12 12:04:24| 55037|[FF D8 FF E0 00 1...|
|file:/Users/jules...|2020-02-12 12:04:24| 54634|[FF D8 FF E0 00 1...|
|file:/Users/jules...|2020-02-12 12:04:24| 54624|[FF D8 FF E0 00 1...|
|file:/Users/jules...|2020-02-12 12:04:24| 54505|[FF D8 FF E0 00 1...|
|file:/Users/jules...|2020-02-12 12:04:24| 54475|[FF D8 FF E0 00 1...|
+--------------------+-------------------+------+--------------------+
only showing top 5 rows
```

Note that the label column is absent when the recursiveFileLookup option is set to "true".

Currently, the binary file data source does not support writing a DataFrame back to the original file format.

In this section, you got a tour of how to read data into a DataFrame from a range of supported file formats. We also showed you how to create temporary views and tables from the existing built-in data sources. Whether you're using the DataFrame API or SQL, the queries produce identical outcomes. You can examine some of these queries in the notebook available in the GitHub repo (*https://github.com/databricks/Learning SparkV2*) for this book.

Summary

To recap, this chapter explored the interoperability between the DataFrame API and Spark SQL. In particular, you got a flavor of how to use Spark SQL to:

- Create managed and unmanaged tables using Spark SQL and the DataFrame API.

- Read from and write to various built-in data sources and file formats.

- Employ the spark.sql programmatic interface to issue SQL queries on structured data stored as Spark SQL tables or views.

- Peruse the Spark `Catalog` to inspect metadata associated with tables and views.

- Use the `DataFrameWriter` and `DataFrameReader` APIs.

Through the code snippets in the chapter and the notebooks available in the book's GitHub repo (*https://github.com/databricks/LearningSparkV2*), you got a feel for how to use DataFrames and Spark SQL. Continuing in this vein, the next chapter further explores how Spark interacts with the external data sources shown in Figure 4-1. You'll see some more in-depth examples of transformations and the interoperability between the DataFrame API and Spark SQL.

Spark SQL and DataFrames: Interacting with External Data Sources

In the previous chapter, we explored interacting with the built-in data sources in Spark. We also took a closer look at the DataFrame API and its interoperability with Spark SQL. In this chapter, we will focus on how Spark SQL interfaces with external components. Specifically, we discuss how Spark SQL allows you to:

- Use user-defined functions for both Apache Hive and Apache Spark.
- Connect with external data sources such as JDBC and SQL databases, PostgreSQL, MySQL, Tableau, Azure Cosmos DB, and MS SQL Server.
- Work with simple and complex types, higher-order functions, and common relational operators.

We'll also look at some different options for querying Spark using Spark SQL, such as the Spark SQL shell, Beeline, and Tableau.

Spark SQL and Apache Hive

Spark SQL is a foundational component of Apache Spark that integrates relational processing with Spark's functional programming API. Its genesis was in previous work on Shark (*https://oreil.ly/QEixA*). Shark was originally built on the Hive codebase on top of Apache Spark[1] and became one of the first interactive SQL query engines on Hadoop systems. It demonstrated that it was possible to have the best of

---

1 The current Spark SQL engine no longer uses the Hive code in its implementation.

both worlds (*https://oreil.ly/FrPY6*); as fast as an enterprise data warehouse, and scaling as well as Hive/MapReduce.

Spark SQL lets Spark programmers leverage the benefits of faster performance and relational programming (e.g., declarative queries and optimized storage), as well as call complex analytics libraries (e.g., machine learning). As discussed in the previous chapter, as of Apache Spark 2.x, the `SparkSession` provides a single unified entry point to manipulate data in Spark.

User-Defined Functions

While Apache Spark has a plethora of built-in functions, the flexibility of Spark allows for data engineers and data scientists to define their own functions too. These are known as *user-defined functions* (UDFs).

Spark SQL UDFs

The benefit of creating your own PySpark or Scala UDFs is that you (and others) will be able to make use of them within Spark SQL itself. For example, a data scientist can wrap an ML model within a UDF so that a data analyst can query its predictions in Spark SQL without necessarily understanding the internals of the model.

Here's a simplified example of creating a Spark SQL UDF. Note that UDFs operate per session and they will not be persisted in the underlying metastore:

```
// In Scala
// Create cubed function
val cubed = (s: Long) => {
  s * s * s
}

// Register UDF
spark.udf.register("cubed", cubed)

// Create temporary view
spark.range(1, 9).createOrReplaceTempView("udf_test")

# In Python
from pyspark.sql.types import LongType

# Create cubed function
def cubed(s):
  return s * s * s

# Register UDF
spark.udf.register("cubed", cubed, LongType())

# Generate temporary view
spark.range(1, 9).createOrReplaceTempView("udf_test")
```

You can now use Spark SQL to execute either of these cubed() functions:

```
// In Scala/Python
// Query the cubed UDF
spark.sql("SELECT id, cubed(id) AS id_cubed FROM udf_test").show()
```

```
+---+--------+
| id|id_cubed|
+---+--------+
|  1|       1|
|  2|       8|
|  3|      27|
|  4|      64|
|  5|     125|
|  6|     216|
|  7|     343|
|  8|     512|
+---+--------+
```

Evaluation order and null checking in Spark SQL

Spark SQL (this includes SQL, the DataFrame API, and the Dataset API) does not guarantee the order of evaluation of subexpressions. For example, the following query does not guarantee that the s is NOT NULL clause is executed prior to the strlen(s) > 1 clause:

```
spark.sql("SELECT s FROM test1 WHERE s IS NOT NULL AND strlen(s) > 1")
```

Therefore, to perform proper null checking, it is recommended that you do the following:

1. Make the UDF itself null-aware and do null checking inside the UDF.

2. Use IF or CASE WHEN expressions to do the null check and invoke the UDF in a conditional branch.

Speeding up and distributing PySpark UDFs with Pandas UDFs

One of the previous prevailing issues with using PySpark UDFs was that they had slower performance than Scala UDFs. This was because the PySpark UDFs required data movement between the JVM and Python, which was quite expensive. To resolve this problem, Pandas UDFs (*https://oreil.ly/jo7kl*) (also known as vectorized UDFs) were introduced as part of Apache Spark 2.3. A Pandas UDF uses Apache Arrow to transfer data and Pandas to work with the data. You define a Pandas UDF using the keyword pandas_udf as the decorator, or to wrap the function itself. Once the data is in Apache Arrow format (*https://oreil.ly/TCsur*), there is no longer the need to serialize/pickle the data as it is already in a format consumable by the Python process. Instead of operating on individual inputs row by row, you are operating on a Pandas Series or DataFrame (i.e., vectorized execution).

From Apache Spark 3.0 with Python 3.6 and above, Pandas UDFs were split into two API categories (*https://oreil.ly/rXX-L*): Pandas UDFs and Pandas Function APIs.

Pandas UDFs

With Apache Spark 3.0, Pandas UDFs infer the Pandas UDF type from Python type hints in Pandas UDFs such as `pandas.Series`, `pandas.DataFrame`, `Tuple`, and `Iterator`. Previously you needed to manually define and specify each Pandas UDF type. Currently, the supported cases of Python type hints in Pandas UDFs are Series to Series, Iterator of Series to Iterator of Series, Iterator of Multiple Series to Iterator of Series, and Series to Scalar (a single value).

Pandas Function APIs

Pandas Function APIs allow you to directly apply a local Python function to a PySpark DataFrame where both the input and output are Pandas instances. For Spark 3.0, the supported Pandas Function APIs are grouped map, map, co-grouped map.

For more information, refer to "Redesigned Pandas UDFs with Python Type Hints" on page 354 in Chapter 12.

The following is an example of a scalar Pandas UDF for Spark 3.0:[2]

```
# In Python
# Import pandas
import pandas as pd

# Import various pyspark SQL functions including pandas_udf
from pyspark.sql.functions import col, pandas_udf
from pyspark.sql.types import LongType

# Declare the cubed function
def cubed(a: pd.Series) -> pd.Series:
    return a * a * a

# Create the pandas UDF for the cubed function
cubed_udf = pandas_udf(cubed, returnType=LongType())
```

The preceding code snippet declares a function called `cubed()` that performs a *cubed* operation. This is a regular Pandas function with the additional `cubed_udf = pandas_udf()` call to create our Pandas UDF.

---

2 Note there are slight differences when working with Pandas UDFs between Spark 2.3 (*https://oreil.ly/pIZk-*), 2.4 (*https://oreil.ly/0NYG-*), and 3.0 (*https://oreil.ly/9wA4s*).

Let's start with a simple Pandas Series (as defined for x) and then apply the local function cubed() for the cubed calculation:

```
# Create a Pandas Series
x = pd.Series([1, 2, 3])

# The function for a pandas_udf executed with local Pandas data
print(cubed(x))
```

The output is as follows:

```
0     1
1     8
2    27
dtype: int64
```

Now let's switch to a Spark DataFrame. We can execute this function as a Spark vectorized UDF as follows:

```
# Create a Spark DataFrame, 'spark' is an existing SparkSession
df = spark.range(1, 4)

# Execute function as a Spark vectorized UDF
df.select("id", cubed_udf(col("id"))).show()
```

Here's the output:

```
+---+----------+
| id|cubed(id)|
+---+----------+
|  1|        1|
|  2|        8|
|  3|       27|
+---+----------+
```

As opposed to a local function, using a vectorized UDF will result in the execution of Spark jobs; the previous local function is a Pandas function executed only on the Spark driver. This becomes more apparent when viewing the Spark UI for one of the stages of this pandas_udf function (Figure 5-1).

 For a deeper dive into Pandas UDFs, refer to pandas user-defined functions documentation (*https://oreil.ly/Qi-pb*).

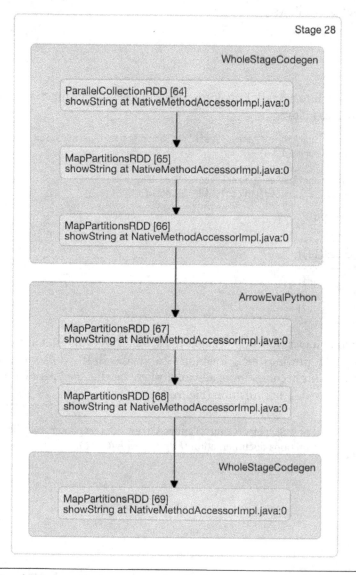

Figure 5-1. Spark UI stages for executing a Pandas UDF on a Spark DataFrame

Like many Spark jobs, the job starts with `parallelize()` to send local data (Arrow binary batches) to executors and calls `mapPartitions()` to convert the Arrow binary batches to Spark's internal data format, which can be distributed to the Spark workers. There are a number of `WholeStageCodegen` steps, which represent a fundamental step up in performance (thanks to Project Tungsten's whole-stage code generation (*https://oreil.ly/5Khvp*), which significantly improves CPU efficiency and performance). But it is the `ArrowEvalPython` step that identifies that (in this case) a Pandas UDF is being executed.

Querying with the Spark SQL Shell, Beeline, and Tableau

There are various mechanisms to query Apache Spark, including the Spark SQL shell, the Beeline CLI utility, and reporting tools like Tableau and Power BI.

In this section, we include instructions for Tableau; for Power BI, please refer to the documentation (*https://oreil.ly/n_KRU*).

Using the Spark SQL Shell

A convenient tool for executing Spark SQL queries is the `spark-sql` CLI. While this utility communicates with the Hive metastore service in local mode, it does not talk to the Thrift JDBC/ODBC server (*https://oreil.ly/kdfko*) (a.k.a. *Spark Thrift Server* or *STS*). The STS allows JDBC/ODBC clients to execute SQL queries over JDBC and ODBC protocols on Apache Spark.

To start the Spark SQL CLI, execute the following command in the $SPARK_HOME folder:

```
./bin/spark-sql
```

Once you've started the shell, you can use it to interactively perform Spark SQL queries. Let's take a look at a few examples.

Create a table

To create a new permanent Spark SQL table, execute the following statement:

```
spark-sql> CREATE TABLE people (name STRING, age int);
```

Your output should be similar to this, noting the creation of the Spark SQL table `people` as well as its file location (`/user/hive/warehouse/people`):

```
20/01/11 22:42:16 WARN HiveMetaStore: Location: file:/user/hive/warehouse/people
specified for non-external table:people
Time taken: 0.63 seconds
```

Insert data into the table

You can insert data into a Spark SQL table by executing a statement similar to:

```
INSERT INTO people SELECT name, age FROM ...
```

As you're not dependent on loading data from a preexisting table or file, you can insert data into the table using INSERT...VALUES statements. These three statements insert three individuals (their names and ages, if known) into the people table:

```
spark-sql> INSERT INTO people VALUES ("Michael", NULL);
Time taken: 1.696 seconds
spark-sql> INSERT INTO people VALUES ("Andy", 30);
Time taken: 0.744 seconds
spark-sql> INSERT INTO people VALUES ("Samantha", 19);
Time taken: 0.637 seconds
spark-sql>
```

Running a Spark SQL query

Now that you have data in your table, you can run Spark SQL queries against it. Let's start by viewing what tables exist in our metastore:

```
spark-sql> SHOW TABLES;
default    people      false
Time taken: 0.016 seconds, Fetched 1 row(s)
```

Next, let's find out how many people in our table are younger than 20 years of age:

```
spark-sql> SELECT * FROM people WHERE age < 20;
Samantha  19
Time taken: 0.593 seconds, Fetched 1 row(s)
```

As well, let's see who the individuals are who did not specify their age:

```
spark-sql> SELECT name FROM people WHERE age IS NULL;
Michael
Time taken: 0.272 seconds, Fetched 1 row(s)
```

Working with Beeline

If you've worked with Apache Hive you may be familiar with the command-line tool Beeline (*https://oreil.ly/Lcrs-*), a common utility for running HiveQL queries against HiveServer2. Beeline is a JDBC client based on the SQLLine CLI (*http://sqlline.source forge.net*). You can use this same utility to execute Spark SQL queries against the Spark Thrift server. Note that the currently implemented Thrift JDBC/ODBC server corresponds to HiveServer2 in Hive 1.2.1. You can test the JDBC server with the following Beeline script that comes with either Spark or Hive 1.2.1.

Start the Thrift server

To start the Spark Thrift JDBC/ODBC server, execute the following command from the $SPARK_HOME folder:

```
./sbin/start-thriftserver.sh
```

> If you have not already started your Spark driver and worker, execute the following command prior to start-thriftserver.sh:
>
> ```
> ./sbin/start-all.sh
> ```

Connect to the Thrift server via Beeline

To test the Thrift JDBC/ODBC server using Beeline, execute the following command:

```
./bin/beeline
```

Then configure Beeline to connect to the local Thrift server:

```
!connect jdbc:hive2://localhost:10000
```

> By default, Beeline is in *non-secure mode*. Thus, the username is your login (e.g., user@learningspark.org) and the password is blank.

Execute a Spark SQL query with Beeline

From here, you can run a Spark SQL query similar to how you would run a Hive query with Beeline. Here are a few sample queries and their output:

```
0: jdbc:hive2://localhost:10000> SHOW tables;

+-----------+------------+--------------+
| database  | tableName  | isTemporary  |
+-----------+------------+--------------+
| default   | people     | false        |
+-----------+------------+--------------+
1 row selected (0.417 seconds)

0: jdbc:hive2://localhost:10000> SELECT * FROM people;

+-----------+-------+
|   name    |  age  |
+-----------+-------+
| Samantha  | 19    |
| Andy      | 30    |
| Michael   | NULL  |
+-----------+-------+
```

```
3 rows selected (1.512 seconds)

0: jdbc:hive2://localhost:10000>
```

Stop the Thrift server

Once you're done, you can stop the Thrift server with the following command:

```
./sbin/stop-thriftserver.sh
```

Working with Tableau

Similar to running queries through Beeline or the Spark SQL CLI, you can connect your favorite BI tool to Spark SQL via the Thrift JDBC/ODBC server. In this section, we will show you how to connect Tableau Desktop (version 2019.2) to your local Apache Spark instance.

 You will need to have the Tableau's Spark ODBC (*https://oreil.ly/ wIGnw*) driver version 1.2.0 or above already installed. If you have installed (or upgraded to) Tableau 2018.1 or greater, this driver should already be preinstalled.

Start the Thrift server

To start the Spark Thrift JDBC/ODBC server, execute the following command from the $SPARK_HOME folder:

```
./sbin/start-thriftserver.sh
```

 If you have not already started your Spark driver and worker, execute the following command prior to start-thriftserver.sh:

```
./sbin/start-all.sh
```

Start Tableau

If you are starting Tableau for the first time, you will be greeted with a Connect dialog that allows you to connect to a plethora of data sources. By default, the Spark SQL option will not be included in the "To a Server" menu on the left (see Figure 5-2).

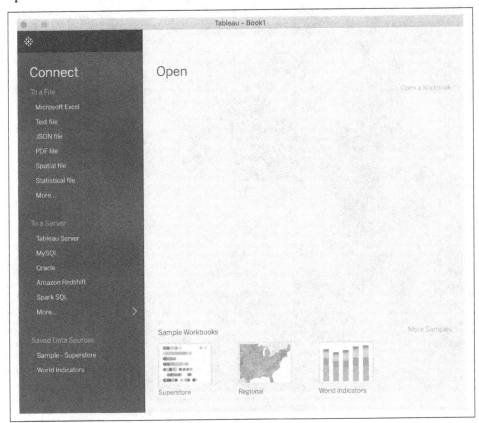

Figure 5-2. Tableau Connect dialog box

To access the Spark SQL option, click More... at the bottom of that list and then choose Spark SQL from the list that appears in the main panel, as shown in Figure 5-3.

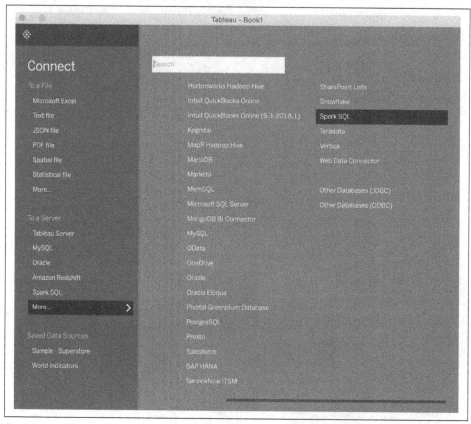

Figure 5-3. Choose More... > Spark SQL to connect to Spark SQL

This will pop up the Spark SQL dialog (Figure 5-4). As you're connecting to a local Apache Spark instance, you can use the non-secure username authentication mode with the following parameters:

- Server: localhost
- Port: 10000 (default)
- Type: SparkThriftServer (default)
- Authentication: Username
- Username: Your login, e.g., user@learningspark.org
- Require SSL: Not checked

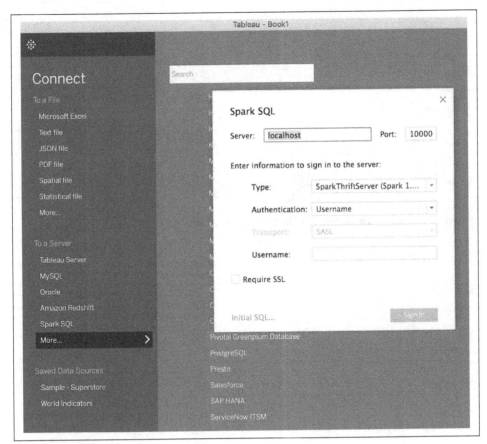

Figure 5-4. The Spark SQL dialog box

Once you have successfully connected to the Spark SQL data source, you will see a Data Source Connections view similar to Figure 5-5.

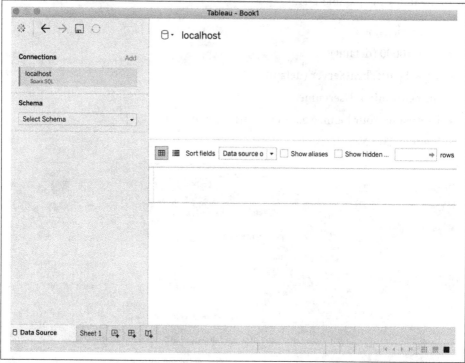

Figure 5-5. Tableau Data Source Connections view, connected to a local Spark instance

From the Select Schema drop-down menu on the left, choose "default." Then enter the name of the table you want to query (see Figure 5-6). Note that you can click the magnifying glass icon to get a full list of the tables that are available.

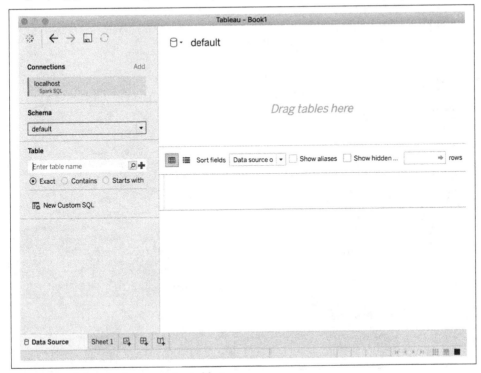

Figure 5-6. Select a schema and a table to query

 For more information on using Tableau to connect to a Spark SQL database, refer to Tableau's Spark SQL documentation (*https://oreil.ly/2A6L7*) and the Databricks Tableau documentation (*https://oreil.ly/--OXu*).

Enter `people` as the table name, then drag and drop the table from the left side into the main dialog (in the space marked "Drag tables here"). You should see something like Figure 5-7.

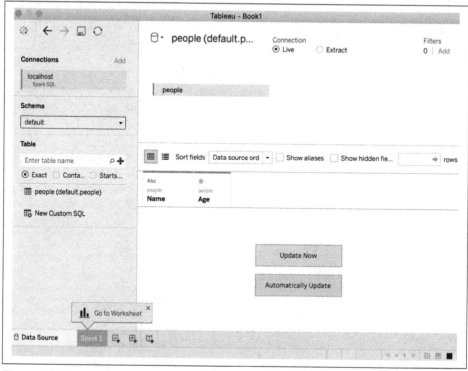

Figure 5-7. Connecting to the people table in your local Spark instance

Click Update Now, and under the covers Tableau will query your Spark SQL data source (Figure 5-8).

You can now execute queries against your Spark data source, join tables, and more, just like with any other Tableau data source.

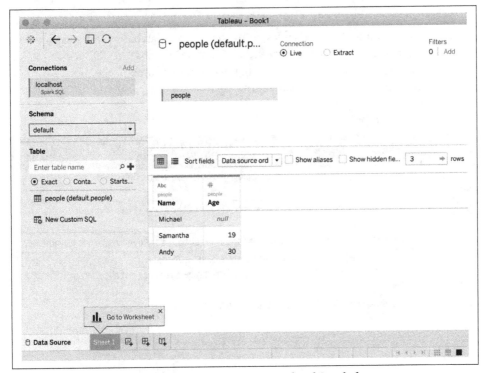

Figure 5-8. Tableau worksheet table view querying a local Spark data source

Stop the Thrift server

Once you're done, you can stop the Thrift server with the following command:

```
./sbin/stop-thriftserver.sh
```

External Data Sources

In this section, we will focus on how to use Spark SQL to connect to external data sources, starting with JDBC and SQL databases.

JDBC and SQL Databases

Spark SQL includes a data source API that can read data from other databases using JDBC (*https://oreil.ly/PHi6y*). It simplifies querying these data sources as it returns the results as a DataFrame, thus providing all of the benefits of Spark SQL (including performance and the ability to join with other data sources).

To get started, you will need to specify the JDBC driver for your JDBC data source and it will need to be on the Spark classpath. From the $SPARK_HOME folder, you'll issue a command like the following:

```
./bin/spark-shell --driver-class-path $database.jar --jars $database.jar
```

Using the data source API, the tables from the remote database can be loaded as a DataFrame or Spark SQL temporary view. Users can specify the JDBC connection properties in the data source options. Table 5-1 contains some of the more common connection properties (case-insensitive) that Spark supports.

Table 5-1. Common connection properties

Property name	Description
user, pass word	These are normally provided as connection properties for logging into the data sources.
url	JDBC connection URL, e.g., `jdbc:postgresql://localhost/test?user=fred&pass word=secret`.
dbtable	JDBC table to read from or write to. You can't specify the `dbtable` and `query` options at the same time.
query	Query to be used to read data from Apache Spark, e.g., `SELECT column1, column2, ..., col umnN FROM [table|subquery]`. You can't specify the `query` and `dbtable` options at the same time.
driver	Class name of the JDBC driver to use to connect to the specified URL.

For the full list of connection properties, see the Spark SQL documentation (*https://oreil.ly/OUG9A*).

The importance of partitioning

When transferring large amounts of data between Spark SQL and a JDBC external source, it is important to partition your data source. All of your data is going through one driver connection, which can saturate and significantly slow down the performance of your extraction, as well as potentially saturate the resources of your source system. While these JDBC properties are optional, for any large-scale operations it is highly recommended to use the properties shown in Table 5-2.

Table 5-2. Partitioning connection properties

Property name	Description
numPartitions	The maximum number of partitions that can be used for parallelism in table reading and writing. This also determines the maximum number of concurrent JDBC connections.
partitionColumn	When reading an external source, `partitionColumn` is the column that is used to determine the partitions; note, `partitionColumn` must be a numeric, date, or timestamp column.
lowerBound	Sets the minimum value of `partitionColumn` for the partition stride.
upperBound	Sets the maximum value of `partitionColumn` for the partition stride.

Let's take a look at an example (*https://oreil.ly/g7Cjc*) to help you understand how these properties work. Suppose we use the following settings:

- numPartitions: 10
- lowerBound: 0
- upperBound: 10000

Then the stride is equal to 1,000, and 10 partitions will be created. This is the equivalent of executing these 10 queries (one for each partition):

- SELECT * FROM table WHERE partitionColumn BETWEEN 0 and 1000
- SELECT * FROM table WHERE partitionColumn BETWEEN 1000 and 2000
- ...
- SELECT * FROM table WHERE partitionColumn BETWEEN 9000 and 10000

While not all-encompassing, the following are some hints to keep in mind when using these properties:

- A good starting point for numPartitions is to use a multiple of the number of Spark workers. For example, if you have four Spark worker nodes, then perhaps start with 4 or 8 partitions. But it is also important to note how well your source system can handle the read requests. For systems that have processing windows, you can maximize the number of concurrent requests to the source system; for systems lacking processing windows (e.g., an OLTP system continuously processing data), you should reduce the number of concurrent requests to prevent saturation of the source system.

- Initially, calculate the lowerBound and upperBound based on the minimum and maximum partitionColumn *actual* values. For example, if you choose {numPartitions:10, lowerBound: 0, upperBound: 10000}, but all of the values are between 2000 and 4000, then only 2 of the 10 queries (one for each partition) will be doing all of the work. In this scenario, a better configuration would be {numPartitions:10, lowerBound: 0, upperBound: 4000}.

- Choose a partitionColumn that can be uniformly distributed to avoid data skew. For example, if the majority of your partitionColumn has the value 2500, with {numPartitions:10, lowerBound: 0, upperBound: 10000} most of the work will be performed by the task requesting the values between 2000 and 3000. Instead, choose a different partitionColumn, or if possible generate a new one (perhaps a hash of multiple columns) to more evenly distribute your partitions.

PostgreSQL

To connect to a PostgreSQL database, build or download the JDBC jar from Maven (*https://oreil.ly/Tg5Z3*) and add it to your classpath. Then start a Spark shell (spark-shell or pyspark), specifying that jar:

```
bin/spark-shell --jars postgresql-42.2.6.jar
```

The following examples show how to load from and save to a PostgreSQL database using the Spark SQL data source API and JDBC in Scala:

```
// In Scala
// Read Option 1: Loading data from a JDBC source using load method
val jdbcDF1 = spark
  .read
  .format("jdbc")
  .option("url", "jdbc:postgresql:[DBSERVER]")
  .option("dbtable", "[SCHEMA].[TABLENAME]")
  .option("user", "[USERNAME]")
  .option("password", "[PASSWORD]")
  .load()

// Read Option 2: Loading data from a JDBC source using jdbc method
// Create connection properties
import java.util.Properties
val cxnProp = new Properties()
cxnProp.put("user", "[USERNAME]")
cxnProp.put("password", "[PASSWORD]")

// Load data using the connection properties
val jdbcDF2 = spark
  .read
  .jdbc("jdbc:postgresql:[DBSERVER]", "[SCHEMA].[TABLENAME]", cxnProp)

// Write Option 1: Saving data to a JDBC source using save method
jdbcDF1
  .write
  .format("jdbc")
  .option("url", "jdbc:postgresql:[DBSERVER]")
  .option("dbtable", "[SCHEMA].[TABLENAME]")
  .option("user", "[USERNAME]")
  .option("password", "[PASSWORD]")
  .save()

// Write Option 2: Saving data to a JDBC source using jdbc method
jdbcDF2.write
  .jdbc(s"jdbc:postgresql:[DBSERVER]", "[SCHEMA].[TABLENAME]", cxnProp)
```

And here's how to do it in PySpark:

```
# In Python
# Read Option 1: Loading data from a JDBC source using load method
jdbcDF1 = (spark
```

```
  .read
  .format("jdbc")
  .option("url", "jdbc:postgresql://[DBSERVER]")
  .option("dbtable", "[SCHEMA].[TABLENAME]")
  .option("user", "[USERNAME]")
  .option("password", "[PASSWORD]")
  .load())

# Read Option 2: Loading data from a JDBC source using jdbc method
jdbcDF2 = (spark
  .read
  .jdbc("jdbc:postgresql://[DBSERVER]", "[SCHEMA].[TABLENAME]",
        properties={"user": "[USERNAME]", "password": "[PASSWORD]"}))

# Write Option 1: Saving data to a JDBC source using save method
(jdbcDF1
  .write
  .format("jdbc")
  .option("url", "jdbc:postgresql://[DBSERVER]")
  .option("dbtable", "[SCHEMA].[TABLENAME]")
  .option("user", "[USERNAME]")
  .option("password", "[PASSWORD]")
  .save())

# Write Option 2: Saving data to a JDBC source using jdbc method
(jdbcDF2
  .write
  .jdbc("jdbc:postgresql:[DBSERVER]", "[SCHEMA].[TABLENAME]",
        properties={"user": "[USERNAME]", "password": "[PASSWORD]"}))
```

MySQL

To connect to a MySQL database, build or download the JDBC jar from Maven
(*https://oreil.ly/c1sAC*) or MySQL (*https://oreil.ly/bH5zb*) (the latter is easier!) and add
it to your classpath. Then start a Spark shell (spark-shell or pyspark), specifying
that jar:

```
bin/spark-shell --jars mysql-connector-java_8.0.16-bin.jar
```

The following examples show how to load data from and save it to a MySQL database
using the Spark SQL data source API and JDBC in Scala:

```
// In Scala
// Loading data from a JDBC source using load
val jdbcDF = spark
  .read
  .format("jdbc")
  .option("url", "jdbc:mysql://[DBSERVER]:3306/[DATABASE]")
  .option("driver", "com.mysql.jdbc.Driver")
  .option("dbtable", "[TABLENAME]")
  .option("user", "[USERNAME]")
  .option("password", "[PASSWORD]")
```

```
  .load()

// Saving data to a JDBC source using save
jdbcDF
  .write
  .format("jdbc")
  .option("url", "jdbc:mysql://[DBSERVER]:3306/[DATABASE]")
  .option("driver", "com.mysql.jdbc.Driver")
  .option("dbtable", "[TABLENAME]")
  .option("user", "[USERNAME]")
  .option("password", "[PASSWORD]")
  .save()
```

And here's how to do it in Python:

```
# In Python
# Loading data from a JDBC source using load
jdbcDF = (spark
  .read
  .format("jdbc")
  .option("url", "jdbc:mysql://[DBSERVER]:3306/[DATABASE]")
  .option("driver", "com.mysql.jdbc.Driver")
  .option("dbtable", "[TABLENAME]")
  .option("user", "[USERNAME]")
  .option("password", "[PASSWORD]")
  .load())

# Saving data to a JDBC source using save
(jdbcDF
  .write
  .format("jdbc")
  .option("url", "jdbc:mysql://[DBSERVER]:3306/[DATABASE]")
  .option("driver", "com.mysql.jdbc.Driver")
  .option("dbtable", "[TABLENAME]")
  .option("user", "[USERNAME]")
  .option("password", "[PASSWORD]")
  .save())
```

Azure Cosmos DB

To connect to an Azure Cosmos DB database, build or download the JDBC jar from
Maven (*https://oreil.ly/vDVQ6*) or GitHub (*https://oreil.ly/dJMx1*) and add it to your
classpath. Then start a Scala or PySpark shell, specifying this jar (note that this example is using Spark 2.4):

```
bin/spark-shell --jars azure-cosmosdb-spark_2.4.0_2.11-1.3.5-uber.jar
```

You also have the option of using --packages to pull the connector from Spark Packages (*https://spark-packages.org*) using its Maven coordinates:

```
export PKG="com.microsoft.azure:azure-cosmosdb-spark_2.4.0_2.11:1.3.5"
bin/spark-shell --packages $PKG
```

The following examples show how to load data from and save it to an Azure Cosmos DB database using the Spark SQL data source API and JDBC in Scala and PySpark. Note that it is common to use the `query_custom` configuration to make use of the various indexes within Cosmos DB:

```scala
// In Scala
// Import necessary libraries
import com.microsoft.azure.cosmosdb.spark.schema._
import com.microsoft.azure.cosmosdb.spark._
import com.microsoft.azure.cosmosdb.spark.config.Config

// Loading data from Azure Cosmos DB
// Configure connection to your collection
val query = "SELECT c.colA, c.coln FROM c WHERE c.origin = 'SEA'"
val readConfig = Config(Map(
  "Endpoint" -> "https://[ACCOUNT].documents.azure.com:443/",
  "Masterkey" -> "[MASTER KEY]",
  "Database" -> "[DATABASE]",
  "PreferredRegions" -> "Central US;East US2;",
  "Collection" -> "[COLLECTION]",
  "SamplingRatio" -> "1.0",
  "query_custom" -> query
))

// Connect via azure-cosmosdb-spark to create Spark DataFrame
val df = spark.read.cosmosDB(readConfig)
df.count

// Saving data to Azure Cosmos DB
// Configure connection to the sink collection
val writeConfig = Config(Map(
  "Endpoint" -> "https://[ACCOUNT].documents.azure.com:443/",
  "Masterkey" -> "[MASTER KEY]",
  "Database" -> "[DATABASE]",
  "PreferredRegions" -> "Central US;East US2;",
  "Collection" -> "[COLLECTION]",
  "WritingBatchSize" -> "100"
))

// Upsert the DataFrame to Azure Cosmos DB
import org.apache.spark.sql.SaveMode
df.write.mode(SaveMode.Overwrite).cosmosDB(writeConfig)
```

```python
# In Python
# Loading data from Azure Cosmos DB
# Read configuration
query = "SELECT c.colA, c.coln FROM c WHERE c.origin = 'SEA'"
readConfig = {
  "Endpoint" : "https://[ACCOUNT].documents.azure.com:443/",
  "Masterkey" : "[MASTER KEY]",
  "Database" : "[DATABASE]",
  "preferredRegions" : "Central US;East US2",
```

```
    "Collection" : "[COLLECTION]",
    "SamplingRatio" : "1.0",
    "schema_samplesize" : "1000",
    "query_pagesize" : "2147483647",
    "query_custom" : query
}

# Connect via azure-cosmosdb-spark to create Spark DataFrame
df = (spark
  .read
  .format("com.microsoft.azure.cosmosdb.spark")
  .options(**readConfig)
  .load())

# Count the number of flights
df.count()

# Saving data to Azure Cosmos DB
# Write configuration
writeConfig = {
 "Endpoint" : "https://[ACCOUNT].documents.azure.com:443/",
 "Masterkey" : "[MASTER KEY]",
 "Database" : "[DATABASE]",
 "Collection" : "[COLLECTION]",
 "Upsert" : "true"
}

# Upsert the DataFrame to Azure Cosmos DB
(df.write
  .format("com.microsoft.azure.cosmosdb.spark")
  .options(**writeConfig)
  .save())
```

For more information, please refer to the Azure Cosmos DB documentation (*https://oreil.ly/OMXBH*).

MS SQL Server

To connect to an MS SQL Server database, download the JDBC jar (*https://oreil.ly/xHkDl*) and add it to your classpath. Then start a Scala or PySpark shell, specifying this jar:

```
bin/spark-shell --jars mssql-jdbc-7.2.2.jre8.jar
```

The following examples show how to load data from and save it to an MS SQL Server database using the Spark SQL data source API and JDBC in Scala and PySpark:

```scala
// In Scala
// Loading data from a JDBC source
// Configure jdbcUrl
val jdbcUrl = "jdbc:sqlserver://[DBSERVER]:1433;database=[DATABASE]"

// Create a Properties() object to hold the parameters.
// Note, you can create the JDBC URL without passing in the
// user/password parameters directly.
val cxnProp = new Properties()
cxnProp.put("user", "[USERNAME]")
cxnProp.put("password", "[PASSWORD]")
cxnProp.put("driver", "com.microsoft.sqlserver.jdbc.SQLServerDriver")

// Load data using the connection properties
val jdbcDF = spark.read.jdbc(jdbcUrl, "[TABLENAME]", cxnProp)

// Saving data to a JDBC source
jdbcDF.write.jdbc(jdbcUrl, "[TABLENAME]", cxnProp)

# In Python
# Configure jdbcUrl
jdbcUrl = "jdbc:sqlserver://[DBSERVER]:1433;database=[DATABASE]"

# Loading data from a JDBC source
jdbcDF = (spark
  .read
  .format("jdbc")
  .option("url", jdbcUrl)
  .option("dbtable", "[TABLENAME]")
  .option("user", "[USERNAME]")
  .option("password", "[PASSWORD]")
  .load())

# Saving data to a JDBC source
(jdbcDF
  .write
  .format("jdbc")
  .option("url", jdbcUrl)
  .option("dbtable", "[TABLENAME]")
  .option("user", "[USERNAME]")
  .option("password", "[PASSWORD]")
  .save())
```

Other External Sources

There are just some of the many external data sources Apache Spark can connect to; other popular data sources include:

- Apache Cassandra (*https://oreil.ly/j8XSa*)
- Snowflake (*https://oreil.ly/NJOii*)
- MongoDB (*https://oreil.ly/MK64A*)

Higher-Order Functions in DataFrames and Spark SQL

Because complex data types are amalgamations of simple data types, it is tempting to manipulate them directly. There are two typical solutions (*https://oreil.ly/JL1UJ*) for manipulating complex data types:

- Exploding the nested structure into individual rows, applying some function, and then re-creating the nested structure
- Building a user-defined function

These approaches have the benefit of allowing you to think of the problem in tabular format. They typically involve (but are not limited to) using utility functions (*https://oreil.ly/gF-0D*) such as get_json_object(), from_json(), to_json(), explode(), and selectExpr().

Let's take a closer look at these two options.

Option 1: Explode and Collect

In this nested SQL statement, we first explode(values), which creates a new row (with the id) for each element (value) within values:

```
-- In SQL
SELECT id, collect_list(value + 1) AS values
FROM  (SELECT id, EXPLODE(values) AS value
         FROM table) x
GROUP BY id
```

While collect_list() returns a list of objects with duplicates, the GROUP BY statement requires shuffle operations, meaning the order of the re-collected array isn't necessarily the same as that of the original array. As values could be any number of dimensions (a really wide and/or really long array) and we're doing a GROUP BY, this approach could be very expensive.

Option 2: User-Defined Function

To perform the same task (adding 1 to each element in values), we can also create a UDF that uses map() to iterate through each element (value) and perform the addition operation:

```
// In Scala
def addOne(values: Seq[Int]): Seq[Int] = {
    values.map(value => value + 1)
}
val plusOneInt = spark.udf.register("plusOneInt", addOne(_: Seq[Int]): Seq[Int])
```

We could then use this UDF in Spark SQL as follows:

```
spark.sql("SELECT id, plusOneInt(values) AS values FROM table").show()
```

While this is better than using `explode()` and `collect_list()` as there won't be any ordering issues, the serialization and deserialization process itself may be expensive. It's also important to note, however, that `collect_list()` may cause executors to experience out-of-memory issues for large data sets, whereas using UDFs would alleviate these issues.

Built-in Functions for Complex Data Types

Instead of using these potentially expensive techniques, you may be able to use some of the built-in functions for complex data types included as part of Apache Spark 2.4 and later. Some of the more common ones are listed in Table 5-3 (array types) and Table 5-4 (map types).

Table 5-3. Array type functions

Function/Description	Query	Output
`array_distinct(array<T>):` `array<T>` Removes duplicates within an array	`SELECT array_distinct(array(1, 2, 3, null, 3));`	`[1,2,3,null]`
`array_intersect(array<T>,` `array<T>): array<T>` Returns the intersection of two arrays without duplicates	`SELECT array_inter` `sect(array(1, 2, 3), array(1, 3, 5));`	`[1,3]`
`array_union(array<T>,` `array<T>): array<T>` Returns the union of two arrays without duplicates	`SELECT array_union(array(1, 2, 3), array(1, 3, 5));`	`[1,2,3,5]`
`array_except(array<T>,` `array<T>): array<T>` Returns elements in `array1` but not in `array2`, without duplicates	`SELECT array_except(array(1, 2, 3), array(1, 3, 5));`	`[2]`
`array_join(array<String>,` `String[, String]): String` Concatenates the elements of an array using a delimiter	`SELECT` `array_join(array('hello', 'world'), ' ');`	`hello world`
`array_max(array<T>): T` Returns the maximum value within the array; null elements are skipped	`SELECT array_max(array(1, 20, null, 3));`	`20`
`array_min(array<T>): T` Returns the minimum value within the array; null elements are skipped	`SELECT array_min(array(1, 20, null, 3));`	`1`

Function/Description	Query	Output
array_position(array<T>, T): Long Returns the (1-based) index of the first element of the given array as a Long	SELECT array_position(array(3, 2, 1), 1);	3
array_remove(array<T>, T): array<T> Removes all elements that are equal to the given element from the given array	SELECT array_remove(array(1, 2, 3, null, 3), 3);	[1,2,null]
arrays_overlap(array<T>, array<T>): array<T> Returns true if array1 contains at least one non-null element also present in array2	SELECT arrays_overlap(array(1, 2, 3), array(3, 4, 5));	true
array_sort(array<T>): array<T> Sorts the input array in ascending order, with null elements placed at the end of the array	SELECT array_sort(array('b', 'd', null, 'c', 'a'));	["a","b","c","d",null]
concat(array<T>, ...): array<T> Concatenates strings, binaries, arrays, etc.	SELECT concat(array(1, 2, 3), array(4, 5), array(6));	[1,2,3,4,5,6]
flatten(array<array<T>>): array<T> Flattens an array of arrays into a single array	SELECT flatten(array(array(1, 2), array(3, 4)));	[1,2,3,4]
array_repeat(T, Int): array<T> Returns an array containing the specified element the specified number of times	SELECT array_repeat('123', 3);	["123","123","123"]
reverse(array<T>): array<T> Returns a reversed string or an array with the reverse order of elements	SELECT reverse(array(2, 1, 4, 3));	[3,4,1,2]
sequence(T, T[, T]): array<T> Generates an array of elements from start to stop (inclusive) by incremental step	SELECT sequence(1, 5); SELECT sequence(5, 1); SELECT sequence(to_date('2018-01-01'), to_date('2018-03-01'), interval 1 month);	[1,2,3,4,5] [5,4,3,2,1] ["2018-01-01", "2018-02-01", "2018-03-01"]
shuffle(array<T>): array<T> Returns a random permutation of the given array	SELECT shuffle(array(1, 20, null, 3));	[null,3,20,1]

Function/Description	Query	Output
slice(array<T>, Int, Int): array<T> Returns a subset of the given array starting from the given index (counting from the end if the index is negative), of the specified length	SELECT slice(array(1, 2, 3, 4), -2, 2);	[3,4]
array_zip(array<T>, array<U>, ...): array<struct<T, U, ...>> Returns a merged array of structs	SELECT arrays_zip(array(1, 2), array(2, 3), array(3, 4));	[{"0":1,"1":2,"2":3}, {"0":2,"1":3,"2":4}]
element_at(array<T>, Int): T / Returns the element of the given array at the given (1-based) index	SELECT element_at(array(1, 2, 3), 2);	2
cardinality(array<T>): Int An alias of size; returns the size of the given array or a map	SELECT cardinality(array('b', 'd', 'c', 'a'));	4

Table 5-4. Map functions

Function/Description	Query	Output
map_form_arrays(array<K>, array<V>): map<K, V> Creates a map from the given pair of key/value arrays; elements in keys should not be null	SELECT map_from_arrays(array(1.0, 3.0), array('2', '4'));	{"1.0":"2", "3.0":"4"}
map_from_entries(array<struct<K, V>>): map<K, V> Returns a map created from the given array	SELECT map_from_entries(array(struct(1, 'a'), struct(2, 'b')));	{"1":"a", "2":"b"}
map_concat(map<K, V>, ...): map<K, V> Returns the union of the input maps	SELECT map_concat(map(1, 'a', 2, 'b'), map(2, 'c', 3, 'd'));	{"1":"a", "2":"c","3":"d"}
element_at(map<K, V>, K): V Returns the value of the given key, or null if the key is not contained in the map	SELECT element_at(map(1, 'a', 2, 'b'), 2);	b
cardinality(array<T>): Int An alias of size; returns the size of the given array or a map	SELECT cardinality(map(1, 'a', 2, 'b'));	2

Higher-Order Functions

In addition to the previously noted built-in functions, there are higher-order functions that take anonymous lambda functions as arguments. An example of a higher-order function is the following:

```
-- In SQL
transform(values, value -> lambda expression)
```

The transform() function takes an array (values) and anonymous function (lambda expression) as input. The function transparently creates a new array by applying the anonymous function to each element, and then assigning the result to the output array (similar to the UDF approach, but more efficiently).

Let's create a sample data set so we can run some examples:

```
# In Python
from pyspark.sql.types import *
schema = StructType([StructField("celsius", ArrayType(IntegerType()))])

t_list = [[35, 36, 32, 30, 40, 42, 38]], [[31, 32, 34, 55, 56]]
t_c = spark.createDataFrame(t_list, schema)
t_c.createOrReplaceTempView("tC")

# Show the DataFrame
t_c.show()
```

```
// In Scala
// Create DataFrame with two rows of two arrays (tempc1, tempc2)
val t1 = Array(35, 36, 32, 30, 40, 42, 38)
val t2 = Array(31, 32, 34, 55, 56)
val tC = Seq(t1, t2).toDF("celsius")
tC.createOrReplaceTempView("tC")

// Show the DataFrame
tC.show()
```

Here's the output:

```
+--------------------+
|             celsius|
+--------------------+
|[35, 36, 32, 30, ...|
|[31, 32, 34, 55, 56]|
+--------------------+
```

With the preceding DataFrame you can run the following higher-order function queries.

transform()

```
transform(array<T>, function<T, U>): array<U>
```

The transform() function produces an array by applying a function to each element of the input array (similar to a map() function):

```
// In Scala/Python
// Calculate Fahrenheit from Celsius for an array of temperatures
spark.sql("""
SELECT celsius,
```

```
  transform(celsius, t -> ((t * 9) div 5) + 32) as fahrenheit
    FROM tC
""").show()
```

```
+--------------------+--------------------+
|             celsius|          fahrenheit|
+--------------------+--------------------+
|[35, 36, 32, 30, ...|[95, 96, 89, 86, ...|
|[31, 32, 34, 55, 56]|[87, 89, 93, 131,...|
+--------------------+--------------------+
```

filter()

```
filter(array<T>, function<T, Boolean>): array<T>
```

The filter() function produces an array consisting of only the elements of the input array for which the Boolean function is true:

```
// In Scala/Python
// Filter temperatures > 38C for array of temperatures
spark.sql("""
SELECT celsius,
 filter(celsius, t -> t > 38) as high
   FROM tC
""").show()
```

```
+--------------------+--------+
|             celsius|    high|
+--------------------+--------+
|[35, 36, 32, 30, ...|[40, 42]|
|[31, 32, 34, 55, 56]|[55, 56]|
+--------------------+--------+
```

exists()

```
exists(array<T>, function<T, V, Boolean>): Boolean
```

The exists() function returns true if the Boolean function holds for any element in the input array:

```
// In Scala/Python
// Is there a temperature of 38C in the array of temperatures
spark.sql("""
SELECT celsius,
       exists(celsius, t -> t = 38) as threshold
   FROM tC
""").show()
```

```
+--------------------+---------+
|             celsius|threshold|
+--------------------+---------+
|[35, 36, 32, 30, ...|     true|
```

```
|[31, 32, 34, 55, 56]|   false|
+--------------------+--------+
```

reduce()

```
reduce(array<T>, B, function<B, T, B>, function<B, R>)
```

The reduce() function reduces the elements of the array to a single value by merging the elements into a buffer B using function<B, T, B> and applying a finishing function<B, R> on the final buffer:

```
// In Scala/Python
// Calculate average temperature and convert to F
spark.sql("""
SELECT celsius,
       reduce(
          celsius,
          0,
          (t, acc) -> t + acc,
          acc -> (acc div size(celsius) * 9 div 5) + 32
       ) as avgFahrenheit
  FROM tC
""").show()
```

```
+--------------------+-------------+
|             celsius|avgFahrenheit|
+--------------------+-------------+
|[35, 36, 32, 30, ...|           96|
|[31, 32, 34, 55, 56]|          105|
+--------------------+-------------+
```

Common DataFrames and Spark SQL Operations

Part of the power of Spark SQL comes from the wide range of DataFrame operations (also known as untyped Dataset operations) it supports. The list of operations is quite extensive and includes:

- Aggregate functions
- Collection functions
- Datetime functions
- Math functions
- Miscellaneous functions
- Non-aggregate functions
- Sorting functions
- String functions

- UDF functions
- Window functions

For the full list, see the Spark SQL documentation (*https://oreil.ly/e1AYA*).

Within this chapter, we will focus on the following common relational operations:

- Unions and joins
- Windowing
- Modifications

To perform these DataFrame operations, we'll first prepare some data. In the following code snippet, we:

1. Import two files and create two DataFrames, one for airport (`airports`) information and one for US flight delays (`departureDelays`).

2. Using `expr()`, convert the `delay` and `distance` columns from STRING to INT.

3. Create a smaller table, `foo`, that we can focus on for our demo examples; it contains only information on three flights originating from Seattle (SEA) to the destination of San Francisco (SFO) for a small time range.

Let's get started:

```scala
// In Scala
import org.apache.spark.sql.functions._

// Set file paths
val delaysPath =
  "/databricks-datasets/learning-spark-v2/flights/departuredelays.csv"
val airportsPath =
  "/databricks-datasets/learning-spark-v2/flights/airport-codes-na.txt"

// Obtain airports data set
val airports = spark.read
  .option("header", "true")
  .option("inferschema", "true")
  .option("delimiter", "\t")
  .csv(airportsPath)
airports.createOrReplaceTempView("airports")

// Obtain departure Delays data set
val delays = spark.read
  .option("header","true")
  .csv(delaysPath)
  .withColumn("delay", expr("CAST(delay as INT) as delay"))
  .withColumn("distance", expr("CAST(distance as INT) as distance"))
delays.createOrReplaceTempView("departureDelays")
```

```
// Create temporary small table
val foo = delays.filter(
  expr("""origin == 'SEA' AND destination == 'SFO' AND
      date like '01010%' AND delay > 0"""))
foo.createOrReplaceTempView("foo")

# In Python
# Set file paths
from pyspark.sql.functions import expr
tripdelaysFilePath =
  "/databricks-datasets/learning-spark-v2/flights/departuredelays.csv"
airportsFilePath =
  "/databricks-datasets/learning-spark-v2/flights/airport-codes-na.txt"

# Obtain airports data set
airports = (spark.read
  .format("csv")
  .options(header="true", inferSchema="true", sep="\t")
  .load(airportsFilePath))

airports.createOrReplaceTempView("airports")

# Obtain departure delays data set
departureDelays = (spark.read
  .format("csv")
  .options(header="true")
  .load(tripdelaysFilePath))

departureDelays = (departureDelays
  .withColumn("delay", expr("CAST(delay as INT) as delay"))
  .withColumn("distance", expr("CAST(distance as INT) as distance")))

departureDelays.createOrReplaceTempView("departureDelays")

# Create temporary small table
foo = (departureDelays
  .filter(expr("""origin == 'SEA' and destination == 'SFO' and
    date like '01010%' and delay > 0""")))
foo.createOrReplaceTempView("foo")
```

The departureDelays DataFrame contains data on >1.3M flights while the foo Data-Frame contains just three rows with information on flights from SEA to SFO for a specific time range, as noted in the following output:

```
// Scala/Python
spark.sql("SELECT * FROM airports LIMIT 10").show()

+-----------+-----+-------+----+
|       City|State|Country|IATA|
+-----------+-----+-------+----+
| Abbotsford|   BC| Canada| YXX|
|   Aberdeen|   SD|    USA| ABR|
```

```
|     Abilene|   TX|    USA| ABI|
|       Akron|   OH|    USA| CAK|
|     Alamosa|   CO|    USA| ALS|
|      Albany|   GA|    USA| ABY|
|      Albany|   NY|    USA| ALB|
|Albuquerque|   NM|    USA| ABQ|
|  Alexandria|   LA|    USA| AEX|
|   Allentown|   PA|    USA| ABE|
+-----------+-----+-------+----+
```

```
spark.sql("SELECT * FROM departureDelays LIMIT 10").show()
```

```
+--------+-----+--------+------+-----------+
|    date|delay|distance|origin|destination|
+--------+-----+--------+------+-----------+
|01011245|    6|     602|   ABE|        ATL|
|01020600|   -8|     369|   ABE|        DTW|
|01021245|   -2|     602|   ABE|        ATL|
|01020605|   -4|     602|   ABE|        ATL|
|01031245|   -4|     602|   ABE|        ATL|
|01030605|    0|     602|   ABE|        ATL|
|01041243|   10|     602|   ABE|        ATL|
|01040605|   28|     602|   ABE|        ATL|
|01051245|   88|     602|   ABE|        ATL|
|01050605|    9|     602|   ABE|        ATL|
+--------+-----+--------+------+-----------+
```

```
spark.sql("SELECT * FROM foo").show()
```

```
+--------+-----+--------+------+-----------+
|    date|delay|distance|origin|destination|
+--------+-----+--------+------+-----------+
|01010710|   31|     590|   SEA|        SFO|
|01010955|  104|     590|   SEA|        SFO|
|01010730|    5|     590|   SEA|        SFO|
+--------+-----+--------+------+-----------+
```

In the following sections, we will execute union, join, and windowing examples with
this data.

Unions

A common pattern within Apache Spark is to union two different DataFrames with
the same schema together. This can be achieved using the union() method:

```
// Scala
// Union two tables
val bar = delays.union(foo)
bar.createOrReplaceTempView("bar")
bar.filter(expr("""origin == 'SEA' AND destination == 'SFO'
AND date LIKE '01010%' AND delay > 0""")).show()
```

```
# In Python
# Union two tables
bar = departureDelays.union(foo)
bar.createOrReplaceTempView("bar")

# Show the union (filtering for SEA and SFO in a specific time range)
bar.filter(expr("""origin == 'SEA' AND destination == 'SFO'
AND date LIKE '01010%' AND delay > 0""")).show()
```

The bar DataFrame is the union of foo with delays. Using the same filtering criteria results in the bar DataFrame, we see a duplication of the foo data, as expected:

```
-- In SQL
spark.sql("""
SELECT *
  FROM bar
 WHERE origin = 'SEA'
   AND destination = 'SFO'
   AND date LIKE '01010%'
   AND delay > 0
""").show()
```

```
+--------+-----+--------+------+-----------+
|    date|delay|distance|origin|destination|
+--------+-----+--------+------+-----------+
|01010710|   31|     590|   SEA|        SFO|
|01010955|  104|     590|   SEA|        SFO|
|01010730|    5|     590|   SEA|        SFO|
|01010710|   31|     590|   SEA|        SFO|
|01010955|  104|     590|   SEA|        SFO|
|01010730|    5|     590|   SEA|        SFO|
+--------+-----+--------+------+-----------+
```

Joins

A common DataFrame operation is to join two DataFrames (or tables) together. By default, a Spark SQL join is an inner join, with the options being inner, cross, outer, full, full_outer, left, left_outer, right, right_outer, left_semi, and left_anti. More information is available in the documentation (https://oreil.ly/CFEhb) (this is applicable to Scala as well as Python).

The following code sample performs the default of an inner join between the airports and foo DataFrames:

```
// In Scala
foo.join(
  airports.as('air),
  $"air.IATA" === $"origin"
).select("City", "State", "date", "delay", "distance", "destination").show()

# In Python
# Join departure delays data (foo) with airport info
```

```
foo.join(
  airports,
  airports.IATA == foo.origin
).select("City", "State", "date", "delay", "distance", "destination").show()

-- In SQL
spark.sql("""
SELECT a.City, a.State, f.date, f.delay, f.distance, f.destination
  FROM foo f
  JOIN airports a
    ON a.IATA = f.origin
""").show()
```

The preceding code allows you to view the date, delay, distance, and destination information from the foo DataFrame joined to the city and state information from the airports DataFrame:

```
+-------+-----+--------+-----+--------+-----------+
|   City|State|    date|delay|distance|destination|
+-------+-----+--------+-----+--------+-----------+
|Seattle|   WA|01010710|   31|     590|        SFO|
|Seattle|   WA|01010955|  104|     590|        SFO|
|Seattle|   WA|01010730|    5|     590|        SFO|
+-------+-----+--------+-----+--------+-----------+
```

Windowing

A window function (*https://oreil.ly/PV7si*) uses values from the rows in a window (a range of input rows) to return a set of values, typically in the form of another row. With window functions, it is possible to operate on a group of rows while still returning a single value for every input row. In this section, we will show how to use the dense_rank() window function; there are many other functions, as noted in Table 5-5.

Table 5-5. Window functions

	SQL	DataFrame API
Ranking functions	rank()	rank()
	dense_rank()	denseRank()
	percent_rank()	percentRank()
	ntile()	ntile()
	row_number()	rowNumber()
Analytic functions	cume_dist()	cumeDist()
	first_value()	firstValue()
	last_value()	lastValue()
	lag()	lag()
	lead()	lead()

Let's start with a review of the `TotalDelays` (calculated by `sum(Delay)`) experienced by flights originating from Seattle (SEA), San Francisco (SFO), and New York City (JFK) and going to a specific set of destination locations, as noted in the following query:

```sql
-- In SQL
DROP TABLE IF EXISTS departureDelaysWindow;

CREATE TABLE departureDelaysWindow AS
SELECT origin, destination, SUM(delay) AS TotalDelays
  FROM departureDelays
 WHERE origin IN ('SEA', 'SFO', 'JFK')
   AND destination IN ('SEA', 'SFO', 'JFK', 'DEN', 'ORD', 'LAX', 'ATL')
 GROUP BY origin, destination;

SELECT * FROM departureDelaysWindow
```

```
+------+-----------+-----------+
|origin|destination|TotalDelays|
+------+-----------+-----------+
|   JFK|        ORD|       5608|
|   SEA|        LAX|       9359|
|   JFK|        SFO|      35619|
|   SFO|        ORD|      27412|
|   JFK|        DEN|       4315|
|   SFO|        DEN|      18688|
|   SFO|        SEA|      17080|
|   SEA|        SFO|      22293|
|   JFK|        ATL|      12141|
|   SFO|        ATL|       5091|
|   SEA|        DEN|      13645|
|   SEA|        ATL|       4535|
|   SEA|        ORD|      10041|
|   JFK|        SEA|       7856|
|   JFK|        LAX|      35755|
|   SFO|        JFK|      24100|
|   SFO|        LAX|      40798|
|   SEA|        JFK|       4667|
+------+-----------+-----------+
```

What if for each of these origin airports you wanted to find the three destinations that experienced the most delays? You could achieve this by running three different queries for each origin and then unioning the results together, like this:

```sql
-- In SQL
SELECT origin, destination, sum(TotalDelays) as sumTotalDelays
FROM departureDelaysWindow
WHERE origin = 'SEA'
GROUP BY origin, destination
ORDER BY sumTotalDelays DESC
LIMIT 3
```

where [ORIGIN] is the three different origin values of JFK, SEA, and SFO.

But a better approach would be to use a window function like dense_rank() to perform the following calculation:

```
-- In SQL
spark.sql("""
SELECT origin, destination, TotalDelays, rank
  FROM (
     SELECT origin, destination, TotalDelays, dense_rank()
       OVER (PARTITION BY origin ORDER BY TotalDelays DESC) as rank
       FROM departureDelaysWindow
  ) t
 WHERE rank <= 3
""").show()
```

```
+------+-----------+-----------+----+
|origin|destination|TotalDelays|rank|
+------+-----------+-----------+----+
|   SEA|        SFO|      22293|   1|
|   SEA|        DEN|      13645|   2|
|   SEA|        ORD|      10041|   3|
|   SFO|        LAX|      40798|   1|
|   SFO|        ORD|      27412|   2|
|   SFO|        JFK|      24100|   3|
|   JFK|        LAX|      35755|   1|
|   JFK|        SFO|      35619|   2|
|   JFK|        ATL|      12141|   3|
+------+-----------+-----------+----+
```

By using the dense_rank() window function, we can quickly ascertain that the destinations with the worst delays for the three origin cities were:

- Seattle (SEA): San Francisco (SFO), Denver (DEN), and Chicago (ORD)
- San Francisco (SFO): Los Angeles (LAX), Chicago (ORD), and New York (JFK)
- New York (JFK): Los Angeles (LAX), San Francisco (SFO), and Atlanta (ATL)

It's important to note that each window grouping needs to fit in a single executor and will get composed into a single partition during execution. Therefore, you need to ensure that your queries are not unbounded (i.e., limit the size of your window).

Modifications

Another common operation is to perform *modifications* to the DataFrame. While DataFrames themselves are immutable, you can modify them through operations that create new, different DataFrames, with different columns, for example. (Recall from earlier chapters that the underlying RDDs are immutable—i.e., they cannot be changed—to ensure there is data lineage for Spark operations.) Let's start with our previous small DataFrame example:

```
// In Scala/Python
foo.show()


--------+-----+--------+------+-----------+
|    date|delay|distance|origin|destination|
+--------+-----+--------+------+-----------+
|01010710|   31|     590|   SEA|        SFO|
|01010955|  104|     590|   SEA|        SFO|
|01010730|    5|     590|   SEA|        SFO|
+--------+-----+--------+------+-----------+
```

Adding new columns

To add a new column to the foo DataFrame, use the withColumn() method:

```
// In Scala
import org.apache.spark.sql.functions.expr
val foo2 = foo.withColumn(
            "status",
            expr("CASE WHEN delay <= 10 THEN 'On-time' ELSE 'Delayed' END")
          )
```

```
# In Python
from pyspark.sql.functions import expr
foo2 = (foo.withColumn(
        "status",
        expr("CASE WHEN delay <= 10 THEN 'On-time' ELSE 'Delayed' END")
      ))
```

The newly created foo2 DataFrame has the contents of the original foo DataFrame plus the additional status column defined by the CASE statement:

```
// In Scala/Python
foo2.show()


+--------+-----+--------+------+-----------+-------+
|    date|delay|distance|origin|destination| status|
+--------+-----+--------+------+-----------+-------+
|01010710|   31|     590|   SEA|        SFO|Delayed|
|01010955|  104|     590|   SEA|        SFO|Delayed|
|01010730|    5|     590|   SEA|        SFO|On-time|
+--------+-----+--------+------+-----------+-------+
```

Dropping columns

To drop a column, use the drop() method. For example, let's remove the delay column as we now have a status column, added in the previous section:

```
// In Scala
val foo3 = foo2.drop("delay")
foo3.show()

# In Python
foo3 = foo2.drop("delay")
foo3.show()
```

```
+--------+--------+------+-----------+-------+
|    date|distance|origin|destination| status|
+--------+--------+------+-----------+-------+
|01010710|     590|   SEA|        SFO|Delayed|
|01010955|     590|   SEA|        SFO|Delayed|
|01010730|     590|   SEA|        SFO|On-time|
+--------+--------+------+-----------+-------+
```

Renaming columns

You can rename a column using the withColumnRenamed() method:

```
// In Scala
val foo4 = foo3.withColumnRenamed("status", "flight_status")
foo4.show()

# In Python
foo4 = foo3.withColumnRenamed("status", "flight_status")
foo4.show()
```

```
+--------+--------+------+-----------+-------------+
|    date|distance|origin|destination|flight_status|
+--------+--------+------+-----------+-------------+
|01010710|     590|   SEA|        SFO|      Delayed|
|01010955|     590|   SEA|        SFO|      Delayed|
|01010730|     590|   SEA|        SFO|      On-time|
+--------+--------+------+-----------+-------------+
```

Pivoting

When working with your data, sometimes you will need to swap the columns for the rows—i.e., *pivot* your data (*https://oreil.ly/XXmqM*). Let's grab some data to demonstrate this concept:

```
-- In SQL
SELECT destination, CAST(SUBSTRING(date, 0, 2) AS int) AS month, delay
  FROM departureDelays
 WHERE origin = 'SEA'
```

```
+-----------+-----+-----+
|destination|month|delay|
+-----------+-----+-----+
|        ORD|    1|   92|
|        JFK|    1|   -7|
|        DFW|    1|   -5|
```

```
|      MIA|    1|   -3|
|      DFW|    1|   -3|
|      DFW|    1|    1|
|      ORD|    1|  -10|
|      DFW|    1|   -6|
|      DFW|    1|   -2|
|      ORD|    1|   -3|
+----------+-----+-----+
only showing top 10 rows
```

Pivoting allows you to place names in the month column (instead of 1 and 2 you can show Jan and Feb, respectively) as well as perform aggregate calculations (in this case average and max) on the delays by destination and month:

```
-- In SQL
SELECT * FROM (
SELECT destination, CAST(SUBSTRING(date, 0, 2) AS int) AS month, delay
  FROM departureDelays WHERE origin = 'SEA'
)
PIVOT (
  CAST(AVG(delay) AS DECIMAL(4, 2)) AS AvgDelay, MAX(delay) AS MaxDelay
  FOR month IN (1 JAN, 2 FEB)
)
ORDER BY destination
```

```
+-----------+------------+------------+------------+------------+
|destination|JAN_AvgDelay|JAN_MaxDelay|FEB_AvgDelay|FEB_MaxDelay|
+-----------+------------+------------+------------+------------+
|        ABQ|       19.86|         316|       11.42|          69|
|        ANC|        4.44|         149|        7.90|         141|
|        ATL|       11.98|         397|        7.73|         145|
|        AUS|        3.48|          50|       -0.21|          18|
|        BOS|        7.84|         110|       14.58|         152|
|        BUR|       -2.03|          56|       -1.89|          78|
|        CLE|       16.00|          27|        null|        null|
|        CLT|        2.53|          41|       12.96|         228|
|        COS|        5.32|          82|       12.18|         203|
|        CVG|       -0.50|           4|        null|        null|
|        DCA|       -1.15|          50|        0.07|          34|
|        DEN|       13.13|         425|       12.95|         625|
|        DFW|        7.95|         247|       12.57|         356|
|        DTW|        9.18|         107|        3.47|          77|
|        EWR|        9.63|         236|        5.20|         212|
|        FAI|        1.84|         160|        4.21|          60|
|        FAT|        1.36|         119|        5.22|         232|
|        FLL|        2.94|          54|        3.50|          40|
|        GEG|        2.28|          63|        2.87|          60|
|        HDN|       -0.44|          27|       -6.50|           0|
+-----------+------------+------------+------------+------------+
only showing top 20 rows
```

Summary

This chapter explored how Spark SQL interfaces with external components. We discussed creating user-defined functions, including Pandas UDFs, and presented some options for executing Spark SQL queries (including the Spark SQL shell, Beeline, and Tableau). We then provided examples of how to use Spark SQL to connect with a variety of external data sources, such as SQL databases, PostgreSQL, MySQL, Tableau, Azure Cosmos DB, MS SQL Server, and others.

We explored Spark's built-in functions for complex data types, and gave some examples of working with higher-order functions. Finally, we discussed some common relational operators and showed how to perform a selection of DataFrame operations.

In the next chapter, we explore how to work with Datasets, the benefits of strongly typed operations, and when and why to use them.

Spark SQL and Datasets

In Chapters 4 and 5, we covered Spark SQL and the DataFrame API. We looked at how to connect to built-in and external data sources, took a peek at the Spark SQL engine, and explored topics such as the interoperability between SQL and Data-Frames, creating and managing views and tables, and advanced DataFrame and SQL transformations.

Although we briefly introduced the Dataset API in Chapter 3, we skimmed over the salient aspects of how Datasets—strongly typed distributed collections—are created, stored, and serialized and deserialized in Spark.

In this chapter, we go under the hood to understand Datasets: we'll explore working with Datasets in Java and Scala, how Spark manages memory to accommodate Dataset constructs as part of the high-level API, and the costs associated with using Datasets.

Single API for Java and Scala

As you may recall from Chapter 3 (Figure 3-1 and Table 3-6), Datasets offer a unified and singular API for strongly typed objects. Among the languages supported by Spark, only Scala and Java are strongly typed; hence, Python and R support only the untyped DataFrame API.

Datasets are domain-specific typed objects that can be operated on in parallel using functional programming or the DSL operators you're familiar with from the Data-Frame API.

Thanks to this singular API, Java developers no longer risk lagging behind. For example, any future interface or behavior changes to Scala's `groupBy()`, `flatMap()`, `map()`, or `filter()` API will be the same for Java too, because it's a singular interface that is common to both implementations.

Scala Case Classes and JavaBeans for Datasets

If you recall from Chapter 3 (Table 3-2), Spark has internal data types, such as `String Type`, `BinaryType`, `IntegerType`, `BooleanType`, and `MapType`, that it uses to map seamlessly to the language-specific data types in Scala and Java during Spark operations. This mapping is done via encoders, which we discuss later in this chapter.

In order to create `Dataset[T]`, where `T` is your typed object in Scala, you need a case class (*https://oreil.ly/06xko*) that defines the object. Using our example data from Chapter 3 (Table 3-1), say we have a JSON file with millions of entries about bloggers writing about Apache Spark in the following format:

```
{id: 1, first: "Jules", last: "Damji", url: "https://tinyurl.1", date:
"1/4/2016", hits: 4535, campaigns: {"twitter", "LinkedIn"}},
...
{id: 87, first: "Brooke", last: "Wenig", url: "https://tinyurl.2", date:
"5/5/2018", hits: 8908, campaigns: {"twitter", "LinkedIn"}}
```

To create a distributed `Dataset[Bloggers]`, we must first define a Scala case class that defines each individual field that comprises a Scala object. This case class serves as a blueprint or schema for the typed object `Bloggers`:

```
// In Scala
case class Bloggers(id:Int, first:String, last:String, url:String, date:String,
hits: Int, campaigns:Array[String])
```

We can now read the file from the data source:

```
val bloggers = "../data/bloggers.json"
val bloggersDS = spark
  .read
  .format("json")
  .option("path", bloggers)
  .load()
  .as[Bloggers]
```

Each row in the resulting distributed data collection is of type `Bloggers`.

Similarly, you can create a JavaBean class of type `Bloggers` in Java and then use encoders to create a `Dataset<Bloggers>`:

```
// In Java
import org.apache.spark.sql.Encoders;
import java.io.Serializable;

public class Bloggers implements Serializable {
    private int id;
    private String first;
```

```
        private String last;
        private String url;
        private String date;
        private int hits;
        private Array[String] campaigns;

    // JavaBean getters and setters
    int getID() { return id; }
    void setID(int i) { id = i; }
    String getFirst() { return first; }
    void setFirst(String f) { first = f; }
    String getLast() { return last; }
    void setLast(String l) { last = l; }
    String getURL() { return url; }
    void setURL (String u) { url = u; }
    String getDate() { return date; }
    Void setDate(String d) { date = d; }
    int getHits() { return hits; }
    void setHits(int h) { hits = h; }

    Array[String] getCampaigns() { return campaigns; }
    void setCampaigns(Array[String] c) { campaigns = c; }
    }

    // Create Encoder
    Encoder<Bloggers> BloggerEncoder = Encoders.bean(Bloggers.class);
    String bloggers = "../bloggers.json"
    Dataset<Bloggers>bloggersDS = spark
      .read
      .format("json")
      .option("path", bloggers)
      .load()
      .as(BloggerEncoder);
```

As you can see, creating Datasets in Scala and Java requires a bit of forethought, as you have to know all the individual column names and types for the rows you are reading. Unlike with DataFrames, where you can optionally let Spark infer the schema, the Dataset API requires that you define your data types ahead of time and that your case class or JavaBean class matches your schema.

The names of the fields in the Scala case class or Java class definition must match the order in the data source. The column names for each row in the data are automatically mapped to the corresponding names in the class and the types are automatically preserved.

You may use an existing Scala case class or JavaBean class if the field names match with your input data. Working with the Dataset API is as easy, concise, and declarative as working with DataFrames. For most of the Dataset's transformations, you can use the same relational operators you've learned about in the previous chapters.

Let's examine some aspects of working with a sample Dataset.

Working with Datasets

One simple and dynamic way to create a sample Dataset is using a `SparkSession` instance. In this scenario, for illustration purposes, we dynamically create a Scala object with three fields: `uid` (unique ID for a user), `uname` (randomly generated username string), and `usage` (minutes of server or service usage).

Creating Sample Data

First, let's generate some sample data:

```scala
// In Scala
import scala.util.Random._
// Our case class for the Dataset
case class Usage(uid:Int, uname:String, usage: Int)
val r = new scala.util.Random(42)
// Create 1000 instances of scala Usage class
// This generates data on the fly
val data = for (i <- 0 to 1000)
  yield (Usage(i, "user-" + r.alphanumeric.take(5).mkString(""),
  r.nextInt(1000)))
// Create a Dataset of Usage typed data
val dsUsage = spark.createDataset(data)
dsUsage.show(10)
```

```
+---+----------+-----+
|uid|     uname|usage|
+---+----------+-----+
|  0|user-Gpi2C|  525|
|  1|user-DgXDi|  502|
|  2|user-M66yO|  170|
|  3|user-xTOn6|  913|
|  4|user-3xGSz|  246|
|  5|user-2aWRN|  727|
|  6|user-EzZY1|   65|
|  7|user-ZlZMZ|  935|
|  8|user-VjxeG|  756|
|  9|user-iqf1P|    3|
+---+----------+-----+
only showing top 10 rows
```

In Java the idea is similar, but we have to use explicit Encoders (in Scala, Spark handles this implicitly):

```java
// In Java
import org.apache.spark.sql.Encoders;
import org.apache.commons.lang3.RandomStringUtils;
import java.io.Serializable;
import java.util.Random;
import java.util.ArrayList;
import java.util.List;

// Create a Java class as a Bean
public class Usage implements Serializable {
    int uid;                    // user id
    String uname;               // username
    int usage;                  // usage

    public Usage(int uid, String uname, int usage) {
        this.uid = uid;
        this.uname = uname;
        this.usage = usage;
    }
    // JavaBean getters and setters
    public int getUid() { return this.uid; }
    public void setUid(int uid) { this.uid = uid; }
    public String getUname() { return this.uname; }
    public void setUname(String uname) { this.uname = uname; }
    public int getUsage() { return this.usage; }
    public void setUsage(int usage) { this.usage = usage; }

    public Usage() {
    }

    public String toString() {
        return "uid: '" + this.uid + "', uame: '" + this.uname + "', 
        usage: '" + this.usage + "'";
    }
}

// Create an explicit Encoder
Encoder<Usage> usageEncoder = Encoders.bean(Usage.class);
Random rand = new Random();
rand.setSeed(42);
List<Usage> data = new ArrayList<Usage>()

// Create 1000 instances of Java Usage class
for (int i = 0; i < 1000; i++) {
  data.add(new Usage(i, "user" +
  RandomStringUtils.randomAlphanumeric(5),
  rand.nextInt(1000));

// Create a Dataset of Usage typed data
Dataset<Usage> dsUsage = spark.createDataset(data, usageEncoder);
```

The generated Dataset between Scala and Java will differ because the random seed algorithm may be different. Hence, your Scala's and Java's query results will differ.

Now that we have our generated Dataset, dsUsage, let's perform some of the common transformations we have done in previous chapters.

Transforming Sample Data

Recall that Datasets are strongly typed collections of domain-specific objects. These objects can be transformed in parallel using functional or relational operations. Examples of these transformations include map(), reduce(), filter(), select(), and aggregate(). As examples of higher-order functions (*https://oreil.ly/KHaqt*), these methods can take lambdas, closures, or functions as arguments and return the results. As such, they lend themselves well to functional programming (*https://oreil.ly/jvWtM*).

Scala is a functional programming language, and more recently lambdas, functional arguments, and closures have been added to Java too. Let's try a couple of higher-order functions in Spark and use functional programming constructs with the sample data we created earlier.

Higher-order functions and functional programming

For a simple example, let's use filter() to return all the users in our dsUsage Dataset whose usage exceeds 900 minutes. One way to do this is to use a functional expression as an argument to the filter() method:

```
// In Scala
import org.apache.spark.sql.functions._
dsUsage
  .filter(d => d.usage > 900)
  .orderBy(desc("usage"))
  .show(5, false)
```

Another way is to define a function and supply that function as an argument to filter():

```
def filterWithUsage(u: Usage) = u.usage > 900
dsUsage.filter(filterWithUsage(_)).orderBy(desc("usage")).show(5)
```

```
+---+----------+-----+
|uid|     uname|usage|
+---+----------+-----+
|561|user-5n2xY|  999|
|113|user-nnAXr|  999|
|605|user-NL6c4|  999|
```

```
|634|user-L0wci|  999|
|805|user-LX27o|  996|
+---+----------+-----+
only showing top 5 rows
```

In the first case we used a lambda expression, {d.usage > 900}, as an argument to the filter() method, whereas in the second case we defined a Scala function, def filterWithUsage(u: Usage) = u.usage > 900. In both cases, the filter() method iterates over each row of the Usage object in the distributed Dataset and applies the expression or executes the function, returning a new Dataset of type Usage for rows where the value of the expression or function is true. (See the Scala documentation (*https://oreil.ly/5yW8d*) for method signature details.)

In Java, the argument to filter() is of type FilterFunction<T> (*https://oreil.ly/ PBNt4*). This can be defined either inline anonymously or with a named function. For this example, we will define our function by name and assign it to the variable f. Applying this function in filter() will return a new Dataset with all the rows for which our filter condition is true:

```
// In Java
// Define a Java filter function
FilterFunction<Usage> f = new FilterFunction<Usage>() {
    public boolean call(Usage u) {
        return (u.usage > 900);
    }
};

// Use filter with our function and order the results in descending order
dsUsage.filter(f).orderBy(col("usage").desc()).show(5);
```

```
+---+----------+-----+
|uid|uname     |usage|
+---+----------+-----+
|67 |user-qCGvZ|997  |
|878|user-J2HUU|994  |
|668|user-pz2Lk|992  |
|750|user-0zWqR|991  |
|242|user-g0kF6|989  |
+---+----------+-----+
only showing top 5 rows
```

Not all lambdas or functional arguments must evaluate to Boolean values; they can return computed values too. Consider this example using the higher-order function map(), where our aim is to find out the usage cost for each user whose usage value is over a certain threshold so we can offer those users a special price per minute.

```
// In Scala
// Use an if-then-else lambda expression and compute a value
dsUsage.map(u => {if (u.usage > 750) u.usage * .15 else u.usage * .50 })
  .show(5, false)
```

```
// Define a function to compute the usage
def computeCostUsage(usage: Int): Double = {
  if (usage > 750) usage * 0.15 else usage * 0.50
}
// Use the function as an argument to map()
dsUsage.map(u => {computeCostUsage(u.usage)}).show(5, false)
+------+
|value |
+------+
|262.5 |
|251.0 |
|85.0  |
|136.95|
|123.0 |
+------+
only showing top 5 rows
```

To use `map()` in Java, you have to define a `MapFunction<T>` (*https://oreil.ly/BP0iY*). This can either be an anonymous class or a defined class that extends `MapFunc tion<T>`. For this example, we use it inline—that is, in the method call itself:

```
// In Java
// Define an inline MapFunction
dsUsage.map((MapFunction<Usage, Double>) u -> {
    if (u.usage > 750)
        return u.usage * 0.15;
    else
        return u.usage * 0.50;
}, Encoders.DOUBLE()).show(5); // We need to explicitly specify the Encoder
+------+
|value |
+------+
|65.0  |
|114.45|
|124.0 |
|132.6 |
|145.5 |
+------+
only showing top 5 rows
```

Though we have computed values for the cost of usage, we don't know which users the computed values are associated with. How do we get this information?

The steps are simple:

1. Create a Scala case class or JavaBean class, `UsageCost`, with an additional field or column named `cost`.

2. Define a function to compute the `cost` and use it in the `map()` method.

Here's what this looks like in Scala:

```scala
// In Scala
// Create a new case class with an additional field, cost
case class UsageCost(uid: Int, uname:String, usage: Int, cost: Double)

// Compute the usage cost with Usage as a parameter
// Return a new object, UsageCost
def computeUserCostUsage(u: Usage): UsageCost = {
  val v = if (u.usage > 750) u.usage * 0.15 else u.usage * 0.50
    UsageCost(u.uid, u.uname, u.usage, v)
}

// Use map() on our original Dataset
dsUsage.map(u => {computeUserCostUsage(u)}).show(5)
```

```
+---+----------+-----+------+
|uid|     uname|usage|  cost|
+---+----------+-----+------+
|  0|user-Gpi2C|  525| 262.5|
|  1|user-DgXDi|  502| 251.0|
|  2|user-M66yO|  170|  85.0|
|  3|user-xTOn6|  913|136.95|
|  4|user-3xGSz|  246| 123.0|
+---+----------+-----+------+
only showing top 5 rows
```

Now we have a transformed Dataset with a new column, cost, computed by the function in our map() transformation, along with all the other columns.

Likewise, in Java, if we want the cost associated with each user we need to define a JavaBean class UsageCost and MapFunction<T>. For the complete JavaBean example, see the book's GitHub repo (*https://github.com/databricks/LearningSparkV2*); for brevity, we will only show the inline MapFunction<T> here:

```java
// In Java
// Get the Encoder for the JavaBean class
Encoder<UsageCost> usageCostEncoder = Encoders.bean(UsageCost.class);

// Apply map() function to our data
dsUsage.map( (MapFunction<Usage, UsageCost>) u -> {
        double v = 0.0;
        if (u.usage > 750) v = u.usage * 0.15; else v = u.usage * 0.50;
        return new UsageCost(u.uid, u.uname,u.usage, v); },
                        usageCostEncoder).show(5);
```

```
+------+---+----------+-----+
|  cost|uid|     uname|usage|
+------+---+----------+-----+
|  65.0|  0|user-xSyzf|  130|
|114.45|  1|user-iOI72|  763|
| 124.0|  2|user-QHRUk|  248|
```

```
|  132.6|  3|user-8GTjo|  884|
|  145.5|  4|user-U4cU1|  970|
+------+---+----------+-----+
only showing top 5 rows
```

There are a few things to observe about using higher-order functions and Datasets:

- We are using typed JVM objects as arguments to functions.

- We are using dot notation (from object-oriented programming) to access individual fields within the typed JVM object, making it easier to read.

- Some of our functions and lambda signatures can be type-safe, ensuring compile-time error detection and instructing Spark what data types to work on, what operations to perform, etc.

- Our code is readable, expressive, and concise, using Java or Scala language features in lambda expressions.

- Spark provides the equivalent of `map()` and `filter()` without higher-order functional constructs in both Java and Scala, so you are not forced to use functional programming with Datasets or DataFrames. Instead, you can simply use conditional DSL operators or SQL expressions: for example, `dsUsage.filter("usage > 900")` or `dsUsage($"usage" > 900)`. (For more on this, see "Costs of Using Datasets" on page 170.)

- For Datasets we use encoders, a mechanism to efficiently convert data between JVM and Spark's internal binary format for its data types (more on that in "Dataset Encoders" on page 168).

 Higher-order functions and functional programming are not unique to Spark Datasets; you can use them with DataFrames too. Recall that a DataFrame is a `Dataset[Row]`, where `Row` is a generic untyped JVM object that can hold different types of fields. The method signature takes expressions or functions that operate on `Row`, meaning that each `Row`'s data type can be input value to the expression or function.

Converting DataFrames to Datasets

For strong type checking of queries and constructs, you can convert DataFrames to Datasets. To convert an existing DataFrame df to a Dataset of type `SomeCaseClass`, simply use the `df.as[SomeCaseClass]` notation. We saw an example of this earlier:

```
// In Scala
val bloggersDS = spark
  .read
  .format("json")
  .option("path", "/data/bloggers/bloggers.json")
  .load()
  .as[Bloggers]
```

`spark.read.format("json")` returns a `DataFrame<Row>`, which in Scala is a type alias for `Dataset[Row]`. Using `.as[Bloggers]` instructs Spark to use encoders, discussed later in this chapter, to serialize/deserialize objects from Spark's internal memory representation to JVM `Bloggers` objects.

Memory Management for Datasets and DataFrames

Spark is an intensive in-memory distributed big data engine, so its efficient use of memory is crucial to its execution speed.[1] Throughout its release history, Spark's usage of memory has significantly evolved (*https://oreil.ly/sL56g*):

- Spark 1.0 used RDD-based Java objects for memory storage, serialization, and deserialization, which was expensive in terms of resources and slow. Also, storage was allocated *on the Java heap,* so you were at the mercy of the JVM's garbage collection (GC) for large data sets.

- Spark 1.x introduced Project Tungsten (*https://oreil.ly/wCQZB*). One of its prominent features was a new internal row-based format to lay out Datasets and DataFrames in off-heap memory, using offsets and pointers. Spark uses an efficient mechanism called *encoders* to serialize and deserialize between the JVM and its internal Tungsten format. Allocating memory off-heap means that Spark is less encumbered by GC.

- Spark 2.x introduced the second-generation Tungsten engine (*https://oreil.ly/hmjz_*), featuring whole-stage code generation and vectorized column-based memory layout. Built on ideas and techniques from modern compilers, this new version also capitalized on modern CPU and cache architectures for fast parallel data access with the "single instruction, multiple data" (SIMD) approach.

---

1 For more details on how Spark manages memory, check out the references provided in the text and the presentations "Apache Spark Memory Management" (*https://oreil.ly/BlR_u*) and "Deep Dive into Project Tungsten Bringing Spark Closer to Bare Metal" (*https://oreil.ly/YuH3a*).

Dataset Encoders

Encoders convert data in off-heap memory from Spark's internal Tungsten format to JVM Java objects. In other words, they serialize and deserialize Dataset objects from Spark's internal format to JVM objects, including primitive data types. For example, an Encoder[T] will convert from Spark's internal Tungsten format to Dataset[T].

Spark has built-in support for automatically generating encoders for primitive types (e.g., string, integer, long), Scala case classes, and JavaBeans. Compared to Java and Kryo serialization and deserialization, Spark encoders are significantly faster (*https://oreil.ly/zz-x9*).

In our earlier Java example, we explicitly created an encoder:

```
Encoder<UsageCost> usageCostEncoder = Encoders.bean(UsageCost.class);
```

However, for Scala, Spark automatically generates the bytecode for these efficient converters. Let's take a peek at Spark's internal Tungsten row-based format.

Spark's Internal Format Versus Java Object Format

Java objects have large overheads—header info, hashcode, Unicode info, etc. Even a simple Java string such as "abcd" takes 48 bytes of storage, instead of the 4 bytes you might expect. Imagine the overhead to create, for example, a MyClass(Int, String, String) object.

Instead of creating JVM-based objects for Datasets or DataFrames, Spark allocates *off-heap* Java memory to lay out their data and employs encoders to convert the data from in-memory representation to JVM object. For example, Figure 6-1 shows how the JVM object MyClass(Int, String, String) would be stored internally.

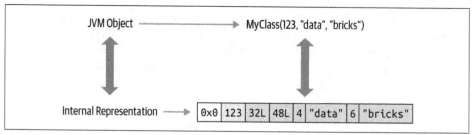

Figure 6-1. JVM object stored in contiguous off-heap Java memory managed by Spark

When data is stored in this contiguous manner and accessible through pointer arithmetic and offets, encoders can quickly serialize or deserialize that data. What does that mean?

Serialization and Deserialization (SerDe)

A concept not new in distributed computing, where data frequently travels over the network among computer nodes in a cluster, *serialization and deserialization* is the process by which a typed object is *encoded* (serialized) into a binary presentation or format by the sender and *decoded* (deserialized) from binary format into its respective data-typed object by the receiver.

For example, if the JVM object `MyClass` in Figure 6-1 had to be shared among nodes in a Spark cluster, the sender would serialize it into an array of bytes, and the receiver would deserialize it back into a JVM object of type `MyClass`.

The JVM has its own built-in Java serializer and deserializer, but it's inefficient because (as we saw in the previous section) the Java objects created by the JVM in the heap memory are bloated. Hence, the process is slow.

This is where the Dataset encoders come to the rescue, for a few reasons:

- Spark's internal Tungsten binary format (see Figures 6-1 and 6-2) stores objects off the Java heap memory, and it's compact so those objects occupy less space.

- Encoders can quickly serialize by traversing across the memory using simple pointer arithmetic with memory addresses and offsets (Figure 6-2).

- On the receiving end, encoders can quickly deserialize the binary representation into Spark's internal representation. Encoders are not hindered by the JVM's garbage collection pauses.

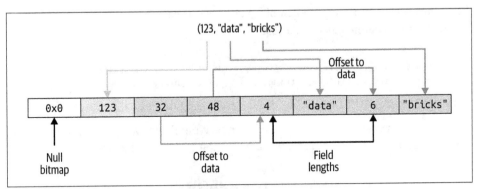

Figure 6-2. Spark's internal Tungsten row-based format

However, most good things in life come at a price, as we discuss next.

Costs of Using Datasets

In "DataFrames Versus Datasets" on page 74 in Chapter 3, we outlined some of the benefits of using Datasets—but these benefits come at a cost. As noted in the preceding section, when Datasets are passed to higher-order functions such as `fil ter()`, `map()`, or `flatMap()` that take lambdas and functional arguments, there is a cost associated with deserializing from Spark's internal Tungsten format into the JVM object.

Compared to other serializers used before encoders were introduced in Spark, this cost is minor and tolerable. However, over larger data sets and many queries, this cost accrues and can affect performance.

Strategies to Mitigate Costs

One strategy to mitigate excessive serialization and deserialization is to use DSL expressions in your queries and avoid excessive use of lambdas as anonymous functions as arguments to higher-order functions. Because lambdas are anonymous and opaque to the Catalyst optimizer until runtime, when you use them it cannot efficiently discern what you're doing (you're not telling Spark *what to do*) and thus cannot optimize your queries (see "The Catalyst Optimizer" on page 77 in Chapter 3).

The second strategy is to chain your queries together in such a way that serialization and deserialization is minimized. Chaining queries together is a common practice in Spark.

Let's illustrate with a simple example. Suppose we have a Dataset of type `Person`, where `Person` is defined as a Scala case class:

```
// In Scala
Person(id: Integer, firstName: String, middleName: String, lastName: String,
    gender: String, birthDate: String, ssn: String, salary: String)
```

We want to issue a set of queries to this Dataset, using functional programming.

Let's examine a case where we compose a query inefficiently, in such a way that we unwittingly incur the cost of repeated serialization and deserialization:

```
import java.util.Calendar
val earliestYear = Calendar.getInstance.get(Calendar.YEAR) - 40

personDS

  // Everyone above 40: lambda-1
  .filter(x => x.birthDate.split("-")(0).toInt > earliestYear)

  // Everyone earning more than 80K
  .filter($"salary" > 80000)
```

```
// Last name starts with J: lambda-2
.filter(x => x.lastName.startsWith("J"))

// First name starts with D
.filter($"firstName".startsWith("D"))
.count()
```

As you can observe in Figure 6-3, each time we move from lambda to DSL (`filter($"salary" > 8000)`) we incur the cost of serializing and deserializing the `Person` JVM object.

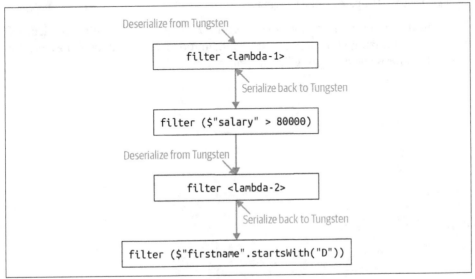

Figure 6-3. An inefficient way to chain queries with lambdas and DSL

By contrast, the following query uses only DSL and no lambdas. As a result, it's much more efficient—no serialization/deserialization is required for the entire composed and chained query:

```
personDS
  .filter(year($"birthDate") > earliestYear) // Everyone above 40
  .filter($"salary" > 80000) // Everyone earning more than 80K
  .filter($"lastName".startsWith("J")) // Last name starts with J
  .filter($"firstName".startsWith("D")) // First name starts with D
  .count()
```

For the curious, you can see the timing difference between the two runs in the notebook for this chapter in the book's GitHub repo (*https://github.com/databricks/LearningSparkV2*).

Summary

In this chapter, we elaborated on how to work with Datasets in Java and Scala. We explored how Spark manages memory to accommodate Dataset constructs as part of its unified and high-level API, and we considered some of the costs associated with using Datasets and how to mitigate those costs. We also showed you how to use Java and Scala's functional programming constructs in Spark.

Finally, we took a look under the hood at how encoders serialize and deserialize from Spark's internal Tungsten binary format to JVM objects.

In the next chapter, we'll look at how to optimize Spark by examining efficient I/O strategies, optimizing and tuning Spark configurations, and what attributes and signals to look for while debugging Spark applications.

Optimizing and Tuning Spark Applications

In the previous chapter, we elaborated on how to work with Datasets in Java and Scala. We explored how Spark manages memory to accommodate Dataset constructs as part of its unified and high-level API, and we considered the costs associated with using Datasets and how to mitigate those costs.

Besides mitigating costs, we also want to consider how to optimize and tune Spark. In this chapter, we will discuss a set of Spark configurations that enable optimizations, look at Spark's family of join strategies, and inspect the Spark UI, looking for clues to bad behavior.

Optimizing and Tuning Spark for Efficiency

While Spark has many configurations for tuning (*https://oreil.ly/c7Y2q*), this book will only cover a handful of the most important and commonly tuned configurations. For a comprehensive list grouped by functional themes, you can peruse the documentation (*https://oreil.ly/mif17*).

Viewing and Setting Apache Spark Configurations

There are three ways you can get and set Spark properties. The first is through a set of configuration files. In your deployment's $SPARK_HOME directory (where you installed Spark), there are a number of config files: *conf/spark-defaults.conf.template*, *conf/log4j.properties.template*, and *conf/spark-env.sh.template*. Changing the default values in these files and saving them without the *.template* suffix instructs Spark to use these new values.

Configuration changes in the *conf/spark-defaults.conf* file apply to the Spark cluster and all Spark applications submitted to the cluster.

The second way is to specify Spark configurations directly in your Spark application or on the command line when submitting the application with `spark-submit`, using the `--conf` flag:

```
spark-submit --conf spark.sql.shuffle.partitions=5 --conf
"spark.executor.memory=2g" --class main.scala.chapter7.SparkConfig_7_1 jars/main-
scala-chapter7_2.12-1.0.jar
```

Here's how you would do this in the Spark application itself:

```scala
// In Scala
import org.apache.spark.sql.SparkSession

def printConfigs(session: SparkSession) = {
  // Get conf
  val mconf = session.conf.getAll
  // Print them
  for (k <- mconf.keySet) { println(s"${k} -> ${mconf(k)}\n") }
}

def main(args: Array[String]) {
 // Create a session
 val spark = SparkSession.builder
   .config("spark.sql.shuffle.partitions", 5)
   .config("spark.executor.memory", "2g")
   .master("local[*]")
   .appName("SparkConfig")
   .getOrCreate()

 printConfigs(spark)
 spark.conf.set("spark.sql.shuffle.partitions",
   spark.sparkContext.defaultParallelism)
 println(" ****** Setting Shuffle Partitions to Default Parallelism")
 printConfigs(spark)
}

spark.driver.host -> 10.8.154.34
spark.driver.port -> 55243
spark.app.name -> SparkConfig
spark.executor.id -> driver
spark.master -> local[*]
spark.executor.memory -> 2g
spark.app.id -> local-1580162894307
spark.sql.shuffle.partitions -> 5
```

The third option is through a programmatic interface via the Spark shell. As with everything else in Spark, APIs are the primary method of interaction. Through the SparkSession object, you can access most Spark config settings.

In a Spark REPL, for example, this Scala code shows the Spark configs on a local host where Spark is launched in local mode (for details on the different modes available, see "Deployment modes" on page 12 in Chapter 1):

```scala
// In Scala
// mconf is a Map[String, String]
scala> val mconf = spark.conf.getAll
...
scala> for (k <- mconf.keySet) { println(s"${k} -> ${mconf(k)}\n") }

spark.driver.host -> 10.13.200.101
spark.driver.port -> 65204
spark.repl.class.uri -> spark://10.13.200.101:65204/classes
spark.jars ->
spark.repl.class.outputDir -> /private/var/folders/jz/qg062ynx5v39wwmfxmph5nn...
spark.app.name -> Spark shell
spark.submit.pyFiles ->
spark.ui.showConsoleProgress -> true
spark.executor.id -> driver
spark.submit.deployMode -> client
spark.master -> local[*]
spark.home -> /Users/julesdamji/spark/spark-3.0.0-preview2-bin-hadoop2.7
spark.sql.catalogImplementation -> hive
spark.app.id -> local-1580144503745
```

You can also view only the Spark SQL–specific Spark configs:

```scala
// In Scala
spark.sql("SET -v").select("key", "value").show(5, false)
```

```python
# In Python
spark.sql("SET -v").select("key", "value").show(n=5, truncate=False)
```

```
+-----------------------------------------------------------+-----------+
|key                                                        |value      |
+-----------------------------------------------------------+-----------+
|spark.sql.adaptive.enabled                                 |false      |
|spark.sql.adaptive.nonEmptyPartitionRatioForBroadcastJoin  |0.2        |
|spark.sql.adaptive.shuffle.fetchShuffleBlocksInBatch.enabled|true       |
|spark.sql.adaptive.shuffle.localShuffleReader.enabled      |true       |
|spark.sql.adaptive.shuffle.maxNumPostShufflePartitions     |<undefined>|
+-----------------------------------------------------------+-----------+
only showing top 5 rows
```

Alternatively, you can access Spark's current configuration through the Spark UI's Environment tab, which we discuss later in this chapter, as read-only values, as shown in Figure 7-1.

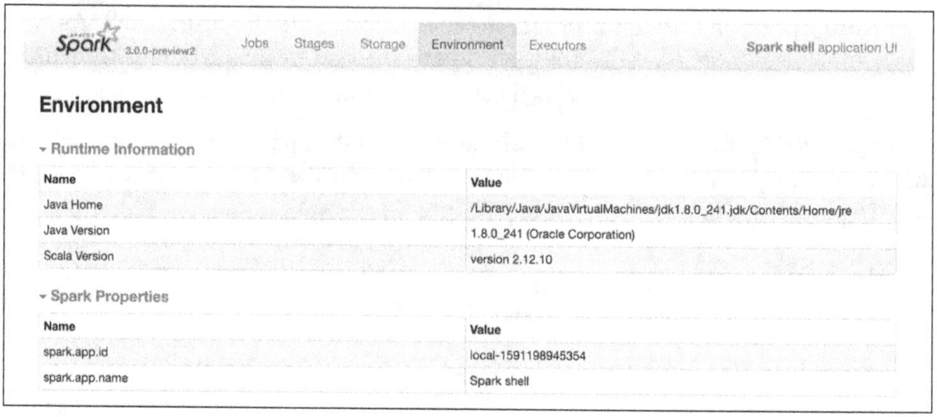

Figure 7-1. The Spark 3.0 UI's Environment tab

To set or modify an existing configuration programmatically, first check if the property is modifiable. `spark.conf.isModifiable("<config_name>")` will return `true` or `false`. All modifiable configs can be set to new values using the API:

```scala
// In Scala
scala> spark.conf.get("spark.sql.shuffle.partitions")
res26: String = 200
scala> spark.conf.set("spark.sql.shuffle.partitions", 5)
scala> spark.conf.get("spark.sql.shuffle.partitions")
res28: String = 5
```

```python
# In Python
>>> spark.conf.get("spark.sql.shuffle.partitions")
'200'
>>> spark.conf.set("spark.sql.shuffle.partitions", 5)
>>> spark.conf.get("spark.sql.shuffle.partitions")
'5'
```

Among all the ways that you can set Spark properties, an order of precedence determines which values are honored. Any values or flags defined in *spark-defaults.conf* will be read first, followed by those supplied on the command line with `spark-submit`, and finally those set via `SparkSession` in the Spark application. All these properties will be merged, with any duplicate properties reset in the Spark application taking precedence. Likewise, values supplied on the command line will supersede settings in the configuration file, provided they are not overwritten in the application itself.

Tweaking or supplying the right configurations helps with performance, as you'll see in the next section. The recommendations here are derived from practitioners' observations in the community and focus on how to maximize cluster resource utilization for Spark to accommodate large-scale workloads.

Scaling Spark for Large Workloads

Large Spark workloads are often batch jobs—some run on a nightly basis, while some are scheduled at regular intervals during the day. In either case, these jobs may process tens of terabytes of data or more. To avoid job failures due to resource starvation or gradual performance degradation, there are a handful of Spark configurations that you can enable or alter. These configurations affect three Spark components: the Spark driver, the executor, and the shuffle service running on the executor.

The Spark driver's responsibility is to coordinate with the cluster manager to launch executors in a cluster and schedule Spark tasks on them. With large workloads, you may have hundreds of tasks. This section explains a few configurations you can tweak or enable to optimize your resource utilization, parallelize tasks, and avoid bottlenecks for large numbers of tasks. Some of the optimization ideas and insights have been derived from big data companies like Facebook that use Spark at terabyte scale, which they shared with the Spark community at the Spark + AI Summit.[1]

Static versus dynamic resource allocation

When you specify compute resources as command-line arguments to `spark-submit`, as we did earlier, you cap the limit. This means that if more resources are needed later as tasks queue up in the driver due to a larger than anticipated workload, Spark cannot accommodate or allocate extra resources.

If instead you use Spark's dynamic resource allocation configuration (*https://oreil.ly/ FX8wl*), the Spark driver can request more or fewer compute resources as the demand of large workloads flows and ebbs. In scenarios where your workloads are dynamic—that is, they vary in their demand for compute capacity—using dynamic allocation helps to accommodate sudden peaks.

One use case where this can be helpful is streaming, where the data flow volume may be uneven. Another is on-demand data analytics, where you might have a high volume of SQL queries during peak hours. Enabling dynamic resource allocation allows Spark to achieve better utilization of resources, freeing executors when not in use and acquiring new ones when needed.

---

1 See "Tuning Apache Spark for Large Scale Workloads" (*https://oreil.ly/cT8Az*) and "Hive Bucketing in Apache Spark" (*https://oreil.ly/S2hTU*).

As well as when working with large or varying workloads, dynamic allocation is also useful in a multitenant environment (*https://oreil.ly/Hqtip*), where Spark may be deployed alongside other applications or services in YARN, Mesos, or Kubernetes. Be advised, however, that Spark's shifting resource demands may impact other applications demanding resources at the same time.

To enable and configure dynamic allocation, you can use settings like the following. Note that the numbers here are arbitrary; the appropriate settings will depend on the nature of your workload and they should be adjusted accordingly. Some of these configs cannot be set inside a Spark REPL, so you will have to set them programmatically:

```
spark.dynamicAllocation.enabled true
spark.dynamicAllocation.minExecutors 2
spark.dynamicAllocation.schedulerBacklogTimeout 1m
spark.dynamicAllocation.maxExecutors 20
spark.dynamicAllocation.executorIdleTimeout 2min
```

By default `spark.dynamicAllocation.enabled` is set to `false`. When enabled with the settings shown here, the Spark driver will request that the cluster manager create two executors to start with, as a minimum (`spark.dynamicAllocation.minExecutors`). As the task queue backlog increases, new executors will be requested each time the backlog timeout (`spark.dynamicAllocation.schedulerBacklogTimeout`) is exceeded. In this case, whenever there are pending tasks that have not been scheduled for over 1 minute, the driver will request that a new executor be launched to schedule backlogged tasks, up to a maximum of 20 (`spark.dynamicAllocation.maxExecutors`). By contrast, if an executor finishes a task and is idle for 2 minutes (`spark.dynamicAllocation.executorIdleTimeout`), the Spark driver will terminate it.

Configuring Spark executors' memory and the shuffle service

Simply enabling dynamic resource allocation is not sufficient. You also have to understand how executor memory is laid out and used by Spark so that executors are not starved of memory or troubled by JVM garbage collection.

The amount of memory available to each executor is controlled by `spark.executor.memory`. This is divided into three sections, as depicted in Figure 7-2: execution memory, storage memory, and reserved memory. The default division is 60% for execution memory and 40% for storage, after allowing for 300 MB for reserved memory, to safeguard against OOM errors. The Spark documentation (*https://oreil.ly/ECABs*) advises that this will work for most cases, but you can adjust what fraction of `spark.executor.memory` you want either section to use as a baseline.

When storage memory is not being used, Spark can acquire it for use in execution memory for execution purposes, and vice versa.

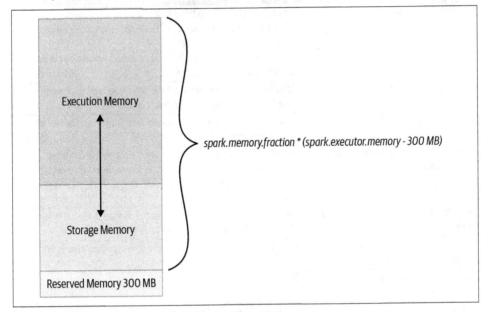

Figure 7-2. Executor memory layout

Execution memory is used for Spark shuffles, joins, sorts, and aggregations. Since different queries may require different amounts of memory, the fraction (`spark.mem ory.fraction` is `0.6` by default) of the available memory to dedicate to this can be tricky to tune but it's easy to adjust. By contrast, storage memory is primarily used for caching user data structures and partitions derived from DataFrames.

During map and shuffle operations, Spark writes to and reads from the local disk's shuffle files, so there is heavy I/O activity. This can result in a bottleneck, because the default configurations are suboptimal for large-scale Spark jobs. Knowing what configurations to tweak can mitigate this risk during this phase of a Spark job.

In Table 7-1, we capture a few recommended configurations to adjust so that the map, spill, and merge processes during these operations are not encumbered by inefficient I/O and to enable these operations to employ buffer memory before writing the final shuffle partitions to disk. Tuning the shuffle service (*https://oreil.ly/4o_pV*) running on each executor can also aid in increasing overall performance for large Spark workloads.

Table 7-1. Spark configurations to tweak for I/O during map and shuffle operations

Configuration	Default value, recommendation, and description
spark.driver.memory	Default is 1g (1 GB). This is the amount of memory allocated to the Spark driver to receive data from executors. This is often changed during spark-submit with --driver-memory. Only change this if you expect the driver to receive large amounts of data back from operations like collect(), or if you run out of driver memory.
spark.shuffle.file.buffer	Default is 32 KB. Recommended is 1 MB. This allows Spark to do more buffering before writing final map results to disk.
spark.file.transferTo	Default is true. Setting it to false will force Spark to use the file buffer to transfer files before finally writing to disk; this will decrease the I/O activity.
spark.shuffle.unsafe.file.out put.buffer	Default is 32 KB. This controls the amount of buffering possible when merging files during shuffle operations. In general, large values (e.g., 1 MB) are more appropriate for larger workloads, whereas the default can work for smaller workloads.
spark.io.compression.lz4.block Size	Default is 32 KB. Increase to 512 KB. You can decrease the size of the shuffle file by increasing the compressed size of the block.
spark.shuffle.service. index.cache.size	Default is 100m. Cache entries are limited to the specified memory footprint in byte.
spark.shuffle.registration. timeout	Default is 5000 ms. Increase to 120000 ms.
spark.shuffle.registration.max Attempts	Default is 3. Increase to 5 if needed.

The recommendations in this table won't work for all situations, but they should give you an idea of how to adjust these configurations based on your workload. Like with everything else in performance tuning, you have to experiment until you find the right balance.

Maximizing Spark parallelism

Much of Spark's efficiency is due to its ability to run multiple tasks in parallel at scale. To understand how you can maximize parallelism—i.e., read and process as much data in parallel as possible—you have to look into how Spark reads data into memory from storage and what partitions mean to Spark.

In data management parlance, a partition is a way to arrange data into a subset of configurable and readable chunks or blocks of contiguous data on disk. These subsets of data can be read or processed independently and in parallel, if necessary, by more than a single thread in a process. This independence matters because it allows for massive parallelism of data processing.

Spark is embarrassingly efficient at processing its tasks in parallel. As you learned in Chapter 2, for large-scale workloads a Spark job will have many stages, and within each stage there will be many tasks. Spark will at best schedule a thread per task per core, and each task will process a distinct partition. To optimize resource utilization and maximize parallelism, the ideal is at least as many partitions as there are cores on the executor, as depicted in Figure 7-3. If there are more partitions than there are cores on each executor, all the cores are kept busy. You can think of partitions as atomic units of parallelism: a single thread running on a single core can work on a single partition.

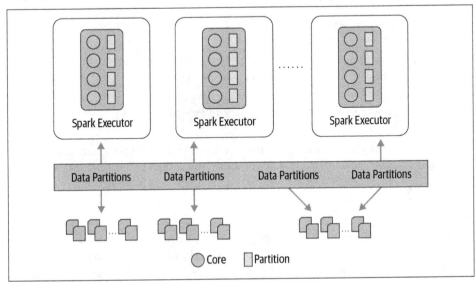

Figure 7-3. Relationship of Spark tasks, cores, partitions, and parallelism

How partitions are created. As mentioned previously, Spark's tasks process data as partitions read from disk into memory. Data on disk is laid out in chunks or contiguous file blocks, depending on the store. By default, file blocks on data stores range in size from 64 MB to 128 MB. For example, on HDFS and S3 the default size is 128 MB (this is configurable). A contiguous collection of these blocks constitutes a partition.

The size of a partition in Spark is dictated by `spark.sql.files.maxPartitionBytes`. The default is 128 MB. You can decrease the size, but that may result in what's known as the "small file problem"—many small partition files, introducing an inordinate amount of disk I/O and performance degradation thanks to filesystem operations such as opening, closing, and listing directories, which on a distributed filesystem can be slow.

Partitions are also created when you explicitly use certain methods of the DataFrame API. For example, while creating a large DataFrame or reading a large file from disk, you can explicitly instruct Spark to create a certain number of partitions:

```
// In Scala
val ds = spark.read.textFile("../README.md").repartition(16)
ds: org.apache.spark.sql.Dataset[String] = [value: string]

ds.rdd.getNumPartitions
res5: Int = 16

val numDF = spark.range(1000L * 1000 * 1000).repartition(16)
numDF.rdd.getNumPartitions

numDF: org.apache.spark.sql.Dataset[Long] = [id: bigint]
res12: Int = 16
```

Finally, *shuffle partitions* are created during the shuffle stage. By default, the number of shuffle partitions is set to 200 in `spark.sql.shuffle.partitions`. You can adjust this number depending on the size of the data set you have, to reduce the amount of small partitions being sent across the network to executors' tasks.

> The default value for `spark.sql.shuffle.partitions` is too high for smaller or streaming workloads; you may want to reduce it to a lower value such as the number of cores on the executors or less.

Created during operations like `groupBy()` or `join()`, also known as wide transformations, shuffle partitions consume both network and disk I/O resources. During these operations, the shuffle will spill results to executors' local disks at the location specified in `spark.local.directory`. Having performant SSD disks for this operation will boost the performance.

There is no magic formula for the number of shuffle partitions to set for the shuffle stage; the number may vary depending on your use case, data set, number of cores, and the amount of executor memory available—it's a trial-and-error approach.[2]

In addition to scaling Spark for large workloads, to boost your performance you'll want to consider caching or persisting your frequently accessed DataFrames or tables. We explore various caching and persistence options in the next section.

---

2 For some tips on configuring shuffle partitions, see "Tuning Apache Spark for Large Scale Workloads" (*https://oreil.ly/QpVyf*), "Hive Bucketing in Apache Spark" (*https://oreil.ly/RmiTd*), and "Why You Should Care about Data Layout in the Filesystem" (*https://oreil.ly/RQQFf*).

Caching and Persistence of Data

What is the difference between caching and persistence? In Spark they are synonymous. Two API calls, `cache()` and `persist()`, offer these capabilities. The latter provides more control over how and where your data is stored—in memory and on disk, serialized and unserialized. Both contribute to better performance for frequently accessed DataFrames or tables.

DataFrame.cache()

`cache()` will store as many of the partitions read in memory across Spark executors as memory allows (see Figure 7-2). While a DataFrame may be fractionally cached, partitions cannot be fractionally cached (e.g., if you have 8 partitions but only 4.5 partitions can fit in memory, only 4 will be cached). However, if not all your partitions are cached, when you want to access the data again, the partitions that are not cached will have to be recomputed, slowing down your Spark job.

Let's look at an example of how caching a large DataFrame improves performance when accessing a DataFrame:

```scala
// In Scala
// Create a DataFrame with 10M records
val df = spark.range(1 * 10000000).toDF("id").withColumn("square", $"id" * $"id")
df.cache() // Cache the data
df.count() // Materialize the cache

res3: Long = 10000000
Command took 5.11 seconds

df.count() // Now get it from the cache
res4: Long = 10000000
Command took 0.44 seconds
```

The first `count()` materializes the cache, whereas the second one accesses the cache, resulting in a close to 12 times faster access time for this data set.

 When you use `cache()` or `persist()`, the DataFrame is not fully cached until you invoke an action that goes through every record (e.g., `count()`). If you use an action like `take(1)`, only one partition will be cached because Catalyst realizes that you do not need to compute all the partitions just to retrieve one record.

Observing how a DataFrame is stored across one executor on a local host, as displayed in Figure 7-4, we can see they all fit in memory (recall that at a low level DataFrames are backed by RDDs).

Figure 7-4. Cache distributed across 12 partitions in executor memory

DataFrame.persist()

persist(StorageLevel.*LEVEL*) is nuanced, providing control over how your data is cached via StorageLevel (*https://oreil.ly/gz6Bb*). Table 7-2 summarizes the different storage levels. Data on disk is always serialized using either Java or Kryo serialization (*https://oreil.ly/NIL6a*).

Table 7-2. StorageLevels

StorageLevel	Description
MEMORY_ONLY	Data is stored directly as objects and stored only in memory.
MEMORY_ONLY_SER	Data is serialized as compact byte array representation and stored only in memory. To use it, it has to be deserialized at a cost.
MEMORY_AND_DISK	Data is stored directly as objects in memory, but if there's insufficient memory the rest is serialized and stored on disk.
DISK_ONLY	Data is serialized and stored on disk.
OFF_HEAP	Data is stored off-heap. Off-heap memory is used in Spark for storage and query execution (*https://oreil.ly/a69L0*); see "Configuring Spark executors' memory and the shuffle service" on page 178.
MEMORY_AND_DISK_SER	Like MEMORY_AND_DISK, but data is serialized when stored in memory. (Data is always serialized when stored on disk.)

> Each StorageLevel (except OFF_HEAP) has an equivalent LEVEL_NAME_2, which means replicate twice on two different Spark executors: MEMORY_ONLY_2, MEMORY_AND_DISK_SER_2, etc. While this option is expensive, it allows data locality in two places, providing fault tolerance and giving Spark the option to schedule a task local to a copy of the data.

Let's look at the same example as in the previous section, but using the persist() method:

```scala
// In Scala
import org.apache.spark.storage.StorageLevel

// Create a DataFrame with 10M records
val df = spark.range(1 * 10000000).toDF("id").withColumn("square", $"id" * $"id")
df.persist(StorageLevel.DISK_ONLY) // Serialize the data and cache it on disk
df.count() // Materialize the cache

res2: Long = 10000000
Command took 2.08 seconds

df.count() // Now get it from the cache
res3: Long = 10000000
Command took 0.38 seconds
```

As you can see from Figure 7-5, the data is persisted on disk, not in memory. To unpersist your cached data, just call DataFrame.unpersist().

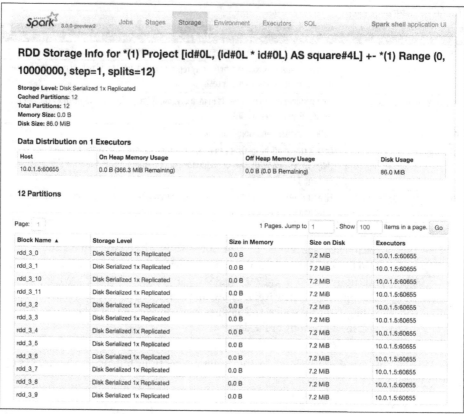

Figure 7-5. Cache distributed across 12 partitions in executor disk

Finally, not only can you cache DataFrames, but you can also cache the tables or views derived from DataFrames. This gives them more readable names in the Spark UI. For example:

```scala
// In Scala
df.createOrReplaceTempView("dfTable")
spark.sql("CACHE TABLE dfTable")
spark.sql("SELECT count(*) FROM dfTable").show()

+--------+
|count(1)|
+--------+
|10000000|
+--------+

Command took 0.56 seconds
```

When to Cache and Persist

Common use cases for caching are scenarios where you will want to access a large data set repeatedly for queries or transformations. Some examples include:

- DataFrames commonly used during iterative machine learning training
- DataFrames accessed commonly for doing frequent transformations during ETL or building data pipelines

When Not to Cache and Persist

Not all use cases dictate the need to cache. Some scenarios that may not warrant caching your DataFrames include:

- DataFrames that are too big to fit in memory
- An inexpensive transformation on a DataFrame not requiring frequent use, regardless of size

As a general rule you should use memory caching judiciously, as it can incur resource costs in serializing and deserializing, depending on the StorageLevel used.

Next, we'll shift our focus to discuss a couple of common Spark join operations that trigger expensive movement of data, demanding compute and network resources from the cluster, and how we can alleviate this movement by organizing the data.

A Family of Spark Joins

Join operations are a common type of transformation in big data analytics in which two data sets, in the form of tables or DataFrames, are merged over a common matching key. Similar to relational databases, the Spark DataFrame and Dataset APIs and Spark SQL offer a series of join transformations: inner joins, outer joins, left joins, right joins, etc. All of these operations trigger a large amount of data movement across Spark executors.

At the heart of these transformations is how Spark computes what data to produce, what keys and associated data to write to the disk, and how to transfer those keys and data to nodes as part of operations like groupBy(), join(), agg(), sortBy(), and reduceByKey(). This movement is commonly referred to as the *shuffle*.

Spark has five distinct join strategies (*https://oreil.ly/q-KvH*) by which it exchanges, moves, sorts, groups, and merges data across executors: the broadcast hash join (BHJ), shuffle hash join (SHJ), shuffle sort merge join (SMJ), broadcast nested loop join (BNLJ), and shuffle-and-replicated nested loop join (a.k.a. Cartesian product join). We'll focus on only two of these here (BHJ and SMJ), because they're the most common ones you'll encounter.

Broadcast Hash Join

Also known as a *map-side-only join*, the broadcast hash join is employed when two data sets, one small (fitting in the driver's and executor's memory) and another large enough to ideally be spared from movement, need to be joined over certain conditions or columns. Using a Spark broadcast variable (*https://oreil.ly/ersei*), the smaller data set is broadcasted by the driver to all Spark executors, as shown in Figure 7-6, and subsequently joined with the larger data set on each executor. This strategy avoids the large exchange.

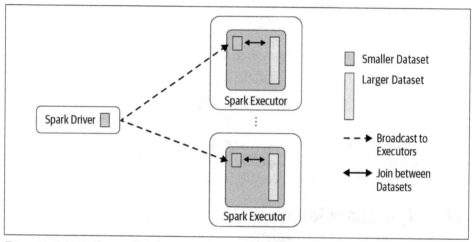

Figure 7-6. BHJ: the smaller data set is broadcast to all executors

By default Spark will use a broadcast join if the smaller data set is less than 10 MB. This configuration is set in `spark.sql.autoBroadcastJoinThreshold`; you can decrease or increase the size depending on how much memory you have on each executor and in the driver. If you are confident that you have enough memory you can use a broadcast join with DataFrames larger than 10 MB (even up to 100 MB).

A common use case is when you have a common set of keys between two Data-Frames, one holding less information than the other, and you need a merged view of both. For example, consider a simple case where you have a large data set of soccer players around the world, `playersDF`, and a smaller data set of soccer clubs they play for, `clubsDF`, and you wish to join them over a common key:

```scala
// In Scala
import org.apache.spark.sql.functions.broadcast
val joinedDF = playersDF.join(broadcast(clubsDF), "key1 === key2")
```

In this code we are forcing Spark to do a broadcast join, but it will resort to this type of join by default if the size of the smaller data set is below the `spark.sql.autoBroadcastJoinThreshold`.

The BHJ is the easiest and fastest join Spark offers, since it does not involve any shuffle of the data set; all the data is available locally to the executor after a broadcast. You just have to be sure that you have enough memory both on the Spark driver's and the executors' side to hold the smaller data set in memory.

At any time after the operation, you can see in the physical plan what join operation was performed by executing:

```
joinedDF.explain(mode)
```

In Spark 3.0, you can use `joinedDF.explain('mode')` to display a readable and digestible output. The modes include `'simple'`, `'extended'`, `'codegen'`, `'cost'`, and `'formatted'`.

When to use a broadcast hash join

Use this type of join under the following conditions for maximum benefit:

- When each key within the smaller and larger data sets is hashed to the same partition by Spark
- When one data set is much smaller than the other (and within the default config of 10 MB, or more if you have sufficient memory)
- When you only want to perform an equi-join, to combine two data sets based on matching unsorted keys
- When you are not worried by excessive network bandwidth usage or OOM errors, because the smaller data set will be broadcast to all Spark executors

Specifying a value of `-1` in `spark.sql.autoBroadcastJoinThreshold` will cause Spark to always resort to a shuffle sort merge join, which we discuss in the next section.

Shuffle Sort Merge Join

The sort-merge algorithm is an efficient way to merge two large data sets over a common key that is sortable, unique, and can be assigned to or stored in the same partition—that is, two data sets with a common hashable key that end up being on the same partition. From Spark's perspective, this means that all rows within each data set with the same key are hashed on the same partition on the same executor. Obviously, this means data has to be colocated or exchanged between executors.

As the name indicates, this join scheme has two phases: a sort phase followed by a merge phase. The sort phase sorts each data set by its desired join key; the merge phase iterates over each key in the row from each data set and merges the rows if the two keys match.

By default, the `SortMergeJoin` is enabled via `spark.sql.join.preferSortMerge Join`. Here is a code snippet from a notebook of standalone applications available for this chapter in the book's GitHub repo (*https://github.com/databricks/Learning SparkV2*). The main idea is to take two large DataFrames, with one million records, and join them on two common keys, `uid == users_id`.

This data is synthetic but illustrates the point:

```scala
// In Scala
import scala.util.Random
// Show preference over other joins for large data sets
// Disable broadcast join
// Generate data
...
spark.conf.set("spark.sql.autoBroadcastJoinThreshold", "-1")

// Generate some sample data for two data sets
var states = scala.collection.mutable.Map[Int, String]()
var items = scala.collection.mutable.Map[Int, String]()
val rnd = new scala.util.Random(42)

// Initialize states and items purchased
states += (0 -> "AZ", 1 -> "CO", 2-> "CA", 3-> "TX", 4 -> "NY", 5-> "MI")
items += (0 -> "SKU-0", 1 -> "SKU-1", 2-> "SKU-2", 3-> "SKU-3", 4 -> "SKU-4",
    5-> "SKU-5")

// Create DataFrames
val usersDF = (0 to 1000000).map(id => (id, s"user_${id}",
    s"user_${id}@databricks.com", states(rnd.nextInt(5))))
    .toDF("uid", "login", "email", "user_state")
val ordersDF = (0 to 1000000)
    .map(r => (r, r, rnd.nextInt(10000), 10 * r* 0.2d,
    states(rnd.nextInt(5)), items(rnd.nextInt(5))))
    .toDF("transaction_id", "quantity", "users_id", "amount", "state", "items")

// Do the join
val usersOrdersDF = ordersDF.join(usersDF, $"users_id" === $"uid")

// Show the joined results
usersOrdersDF.show(false)
```

```
+--------------+--------+--------+--------+-----+-----+---+---+----------+
|transaction_id|quantity|users_id|amount  |state|items|uid|...|user_state|
+--------------+--------+--------+--------+-----+-----+---+---+----------+
|3916          |3916    |148     |7832.0  |CA   |SKU-1|148|...|CO        |
|36384         |36384   |148     |72768.0 |NY   |SKU-2|148|...|CO        |
```

```
|41839          |41839    |148    |83678.0  |CA    |SKU-3|148|...|CO          |
|48212          |48212    |148    |96424.0  |CA    |SKU-4|148|...|CO          |
|48484          |48484    |148    |96968.0  |TX    |SKU-3|148|...|CO          |
|50514          |50514    |148    |101028.0 |CO    |SKU-0|148|...|CO          |
|65694          |65694    |148    |131388.0 |TX    |SKU-4|148|...|CO          |
|65723          |65723    |148    |131446.0 |CA    |SKU-1|148|...|CO          |
|93125          |93125    |148    |186250.0 |NY    |SKU-3|148|...|CO          |
|107097         |107097   |148    |214194.0 |TX    |SKU-2|148|...|CO          |
|111297         |111297   |148    |222594.0 |AZ    |SKU-3|148|...|CO          |
|117195         |117195   |148    |234390.0 |TX    |SKU-4|148|...|CO          |
|253407         |253407   |148    |506814.0 |NY    |SKU-4|148|...|CO          |
|267180         |267180   |148    |534360.0 |AZ    |SKU-0|148|...|CO          |
|283187         |283187   |148    |566374.0 |AZ    |SKU-3|148|...|CO          |
|289245         |289245   |148    |578490.0 |AZ    |SKU-0|148|...|CO          |
|314077         |314077   |148    |628154.0 |CO    |SKU-3|148|...|CO          |
|322170         |322170   |148    |644340.0 |TX    |SKU-3|148|...|CO          |
|344627         |344627   |148    |689254.0 |NY    |SKU-3|148|...|CO          |
|345611         |345611   |148    |691222.0 |TX    |SKU-3|148|...|CO          |
+--------------+--------+--------+--------+-----+-----+---+---+----------+
only showing top 20 rows
```

Examining our final execution plan, we notice that Spark employed a `SortMergeJoin`, as expected, to join the two DataFrames. The `Exchange` operation is the shuffle of the results of the map operation on each executor:

```
usersOrdersDF.explain()

== Physical Plan ==
InMemoryTableScan [transaction_id#40, quantity#41, users_id#42, amount#43,
state#44, items#45, uid#13, login#14, email#15, user_state#16]
   +- InMemoryRelation [transaction_id#40, quantity#41, users_id#42, amount#43,
state#44, items#45, uid#13, login#14, email#15, user_state#16],
StorageLevel(disk, memory, deserialized, 1 replicas)
         +- *(3) SortMergeJoin [users_id#42], [uid#13], Inner
            :- *(1) Sort [users_id#42 ASC NULLS FIRST], false, 0
            :  +- Exchange hashpartitioning(users_id#42, 16), true, [id=#56]
            :     +- LocalTableScan [transaction_id#40, quantity#41, users_id#42,
amount#43, state#44, items#45]
            +- *(2) Sort [uid#13 ASC NULLS FIRST], false, 0
               +- Exchange hashpartitioning(uid#13, 16), true, [id=#57]
                  +- LocalTableScan [uid#13, login#14, email#15, user_state#16]
```

Furthermore, the Spark UI (which we will discuss in the next section) shows three stages for the entire job: the Exchange and Sort operations happen in the final stage, followed by merging of the results, as depicted in Figures 7-7 and 7-8. The Exchange is expensive and requires partitions to be shuffled across the network between executors.

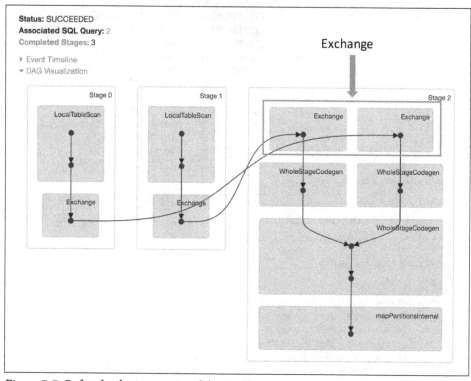

Figure 7-7. Before bucketing: stages of the Spark

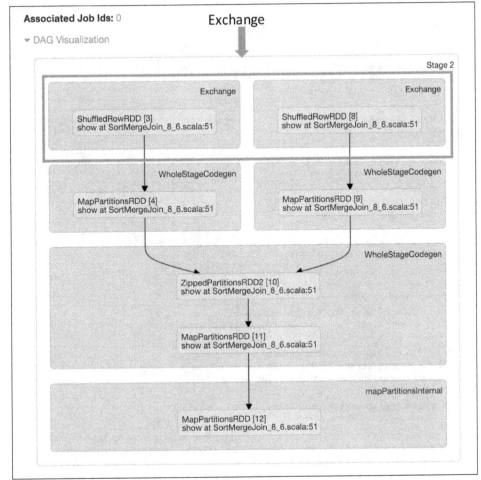

Figure 7-8. Before bucketing: Exchange is required

Optimizing the shuffle sort merge join

We can eliminate the Exchange step from this scheme if we create partitioned buckets for common sorted keys or columns on which we want to perform frequent equi-joins. That is, we can create an explicit number of buckets to store specific sorted columns (one key per bucket). Presorting and reorganizing data in this way boosts performance, as it allows us to skip the expensive Exchange operation and go straight to WholeStageCodegen.

In the following code snippet from the notebook for this chapter (available in the book's GitHub repo (*https://github.com/databricks/LearningSparkV2*)) we sort and bucket by the users_id and uid columns on which we'll join, and save the buckets as Spark managed tables in Parquet format:

```scala
// In Scala
import org.apache.spark.sql.functions._
import org.apache.spark.sql.SaveMode

// Save as managed tables by bucketing them in Parquet format
usersDF.orderBy(asc("uid"))
  .write.format("parquet")
  .bucketBy(8, "uid")
  .mode(SaveMode.OverWrite)
  .saveAsTable("UsersTbl")

ordersDF.orderBy(asc("users_id"))
  .write.format("parquet")
  .bucketBy(8, "users_id")
  .mode(SaveMode.OverWrite)
  .saveAsTable("OrdersTbl")

// Cache the tables
spark.sql("CACHE TABLE UsersTbl")
spark.sql("CACHE TABLE OrdersTbl")

// Read them back in
val usersBucketDF = spark.table("UsersTbl")
val ordersBucketDF = spark.table("OrdersTbl")

// Do the join and show the results
val joinUsersOrdersBucketDF = ordersBucketDF
    .join(usersBucketDF, $"users_id" === $"uid")

joinUsersOrdersBucketDF.show(false)
```

```
+--------------+--------+--------+---------+-----+-----+---+---+----------+
|transaction_id|quantity|users_id|amount   |state|items|uid|...|user_state|
+--------------+--------+--------+---------+-----+-----+---+---+----------+
|144179        |144179  |22      |288358.0 |TX   |SKU-4|22 |...|CO        |
|145352        |145352  |22      |290704.0 |NY   |SKU-0|22 |...|CO        |
|168648        |168648  |22      |337296.0 |TX   |SKU-2|22 |...|CO        |
|173682        |173682  |22      |347364.0 |NY   |SKU-2|22 |...|CO        |
|397577        |397577  |22      |795154.0 |CA   |SKU-3|22 |...|CO        |
|403974        |403974  |22      |807948.0 |CO   |SKU-2|22 |...|CO        |
|405438        |405438  |22      |810876.0 |NY   |SKU-1|22 |...|CO        |
|417886        |417886  |22      |835772.0 |CA   |SKU-3|22 |...|CO        |
|420809        |420809  |22      |841618.0 |NY   |SKU-4|22 |...|CO        |
|659905        |659905  |22      |1319810.0|AZ   |SKU-1|22 |...|CO        |
|899422        |899422  |22      |1798844.0|TX   |SKU-4|22 |...|CO        |
|906616        |906616  |22      |1813232.0|CO   |SKU-2|22 |...|CO        |
|916292        |916292  |22      |1832584.0|TX   |SKU-0|22 |...|CO        |
|916827        |916827  |22      |1833654.0|TX   |SKU-1|22 |...|CO        |
|919106        |919106  |22      |1838212.0|TX   |SKU-1|22 |...|CO        |
|921921        |921921  |22      |1843842.0|AZ   |SKU-4|22 |...|CO        |
|926777        |926777  |22      |1853554.0|CO   |SKU-2|22 |...|CO        |
|124630        |124630  |22      |249260.0 |CO   |SKU-0|22 |...|CO        |
```

```
|129823         |129823  |22       |259646.0 |NY   |SKU-4|22 |...|CO         |
|132756         |132756  |22       |265512.0 |AZ   |SKU-2|22 |...|CO         |
+---------------+--------+---------+---------+-----+-----+---+---+-----------+
only showing top 20 rows
```

The joined output is sorted by `uid` and `users_id`, because we saved the tables sorted in ascending order. As such, there's no need to sort during the `SortMergeJoin`. Looking at the Spark UI (Figure 7-9), we can see that we skipped the `Exchange` and went straight to `WholeStageCodegen`.

The physical plan also shows no `Exchange` was performed, compared to the physical plan before bucketing:

```
joinUsersOrdersBucketDF.explain()

== Physical Plan ==
*(3) SortMergeJoin [users_id#165], [uid#62], Inner
:- *(1) Sort [users_id#165 ASC NULLS FIRST], false, 0
:  +- *(1) Filter isnotnull(users_id#165)
:     +- Scan In-memory table `OrdersTbl` [transaction_id#163, quantity#164,
users_id#165, amount#166, state#167, items#168], [isnotnull(users_id#165)]
:           +- InMemoryRelation [transaction_id#163, quantity#164, users_id#165,
amount#166, state#167, items#168], StorageLevel(disk, memory, deserialized, 1
replicas)
:                 +- *(1) ColumnarToRow
:                    +- FileScan parquet
...
```

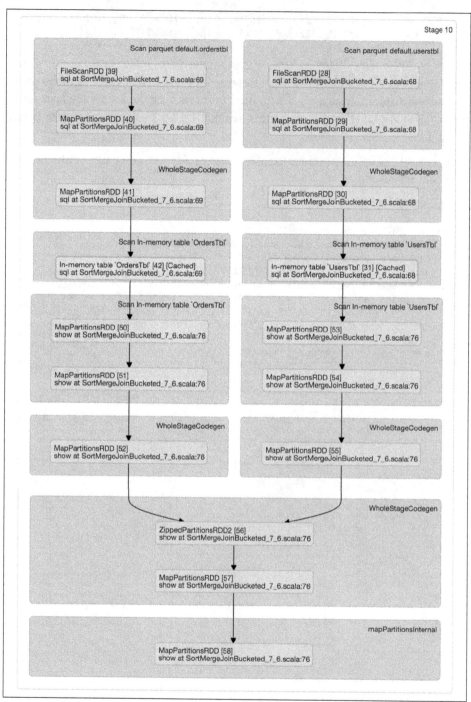

Figure 7-9. After bucketing: Exchange is not required

When to use a shuffle sort merge join

Use this type of join under the following conditions for maximum benefit:

- When each key within two large data sets can be sorted and hashed to the same partition by Spark
- When you want to perform only equi-joins to combine two data sets based on matching sorted keys
- When you want to prevent Exchange and Sort operations to save large shuffles across the network

So far we have covered operational aspects related to tuning and optimizing Spark, and how Spark exchanges data during two common join operations. We also demonstrated how you can boost the performance of a shuffle sort merge join operation by using bucketing to avoid large exchanges of data.

As you've seen in the preceding figures, the Spark UI is a useful way to visualize these operations. It shows collected metrics and the state of the program, revealing a wealth of information and clues about possible performance bottlenecks. In the final section of this chapter, we discuss what to look for in the Spark UI.

Inspecting the Spark UI

Spark provides an elaborate web UI that allows us to inspect various components of our applications. It offers details on memory usage, jobs, stages, and tasks, as well as event timelines, logs, and various metrics and statistics that can give you insight into what transpires in your Spark applications, both at the Spark driver level and in individual executors.

A spark-submit job will launch the Spark UI, and you can connect to it on the local host (in local mode) or through the Spark driver (in other modes) at the default port 4040.

Journey Through the Spark UI Tabs

The Spark UI has six tabs, as shown in Figure 7-10, each providing opportunities for exploration. Let's take a look at what each tab reveals to us.

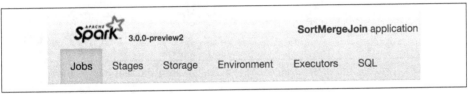

Figure 7-10. Spark UI tabs

This discussion applies to Spark 2.x and Spark 3.0. While much of the UI is the same in Spark 3.0, it also adds a seventh tab, Structured Streaming. This is previewed in Chapter 12.

Jobs and Stages

As you learned in Chapter 2, Spark breaks an application down into jobs, stages, and tasks. The Jobs and Stages tabs allow you to navigate through these and drill down to a granular level to examine the details of individual tasks. You can view their completion status and review metrics related to I/O, memory consumption, duration of execution, etc.

Figure 7-11 shows the Jobs tab with the expanded Event Timeline, showing when executors were added to or removed from the cluster. It also provides a tabular list of all completed jobs in the cluster. The Duration column indicates the time it took for each job (identified by the Job Id in the first column) to finish. If this time is high, it's a good indication that you might want to investigate the stages in that job to see what tasks might be causing delays. From this summary page you can also access a details page for each job, including a DAG visualization and list of completed stages.

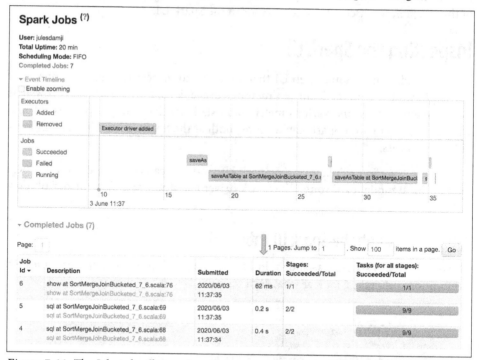

Figure 7-11. The Jobs tab offers a view of the event timeline and list of all completed jobs

The Stages tab provides a summary of the current state of all stages of all jobs in the application. You can also access a details page for each stage, providing a DAG and metrics on its tasks (Figure 7-12). As well as some other optional statistics, you can see the average duration of each task, time spent in garbage collection (GC), and number of shuffle bytes/records read. If shuffle data is being read from remote executors, a high Shuffle Read Blocked Time can signal I/O issues. A high GC time signals too many objects on the heap (your executors may be memory-starved). If a stage's max task time is much larger than the median, then you probably have data skew caused by uneven data distribution in your partitions. Look for these tell-tale signs.

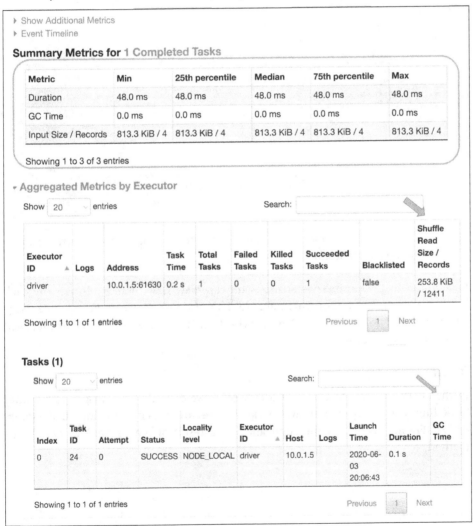

Figure 7-12. The Stages tab provides details on stages and their tasks

You can also see aggregated metrics for each executor and a breakdown of the individual tasks on this page.

Executors

The Executors tab provides information on the executors created for the application. As you can see in Figure 7-13, you can drill down into the minutiae of details about resource usage (disk, memory, cores), time spent in GC, amount of data written and read during shuffle, etc.

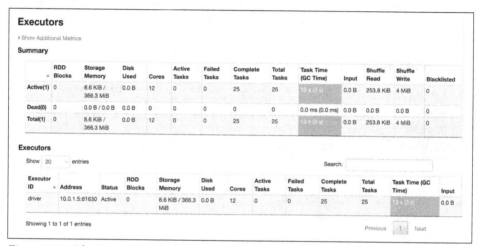

Figure 7-13. The Executors tab shows granular statistics and metrics on the executors used by your Spark application

In addition to the summary statistics, you can view how memory is used by each individual executor, and for what purpose. This also helps to examine resource usage when you have used the cache() or persist() method on a DataFrame or managed table, which we discuss next.

Storage

In the Spark code in "Shuffle Sort Merge Join" we cached two managed tables after bucketing. The Storage tab, shown in Figure 7-14, provides information on any tables or DataFrames cached by the application as a result of the cache() or persist() method.

ID	RDD Name	Storage Level	Cached Partitions	Fraction Cached	Size in Memory	Size on Disk
2	LocalTableScan [uid#13, login#14, email#15, user_state#16]	Disk Serialized 1x Replicated	12	100%	0.0 B	4.4 MiB
16	LocalTableScan [transaction_id#40, quantity#41, users_id#42, amount#43, state#44, items#45]	Disk Serialized 1x Replicated	12	100%	0.0 B	1771.0 KiB
31	In-memory table `UsersTbl`	Disk Memory Deserialized 1x Replicated	8	100%	4.4 MiB	0.0 B
42	In-memory table `OrdersTbl`	Disk Memory Deserialized 1x Replicated	8	100%	2.0 MiB	0.0 B

Figure 7-14. The Storage tab shows details on memory usage

Going a bit further by clicking on the link "In-memory table `UsersTbl`" in Figure 7-14 displays how the table is cached in memory and on disk across 1 executor and 8 partitions—this number corresponds to the number of buckets we created for this table (see Figure 7-15).

RDD Storage Info for In-memory table `UsersTbl`

Storage Level: Disk Memory Deserialized 1x Replicated
Cached Partitions: 8
Total Partitions: 8
Memory Size: 4.4 MiB
Disk Size: 0.0 B

Data Distribution on 1 Executors

Host	On Heap Memory Usage	Off Heap Memory Usage	Disk Usage
10.0.1.5:61782	4.4 MiB (359.8 MiB Remaining)	0.0 B (0.0 B Remaining)	0.0 B

Page: 1   1 Pages. Jump to 1 . Show 100 items in a page. Go

Block Name ▲	Storage Level	Size in Memory	Size on Disk	Executors
rdd_31_0	Memory Deserialized 1x Replicated	558.3 KiB	0.0 B	10.0.1.5:61782
rdd_31_1	Memory Deserialized 1x Replicated	556.3 KiB	0.0 B	10.0.1.5:61782
rdd_31_2	Memory Deserialized 1x Replicated	565.7 KiB	0.0 B	10.0.1.5:61782
rdd_31_3	Memory Deserialized 1x Replicated	560.5 KiB	0.0 B	10.0.1.5:61782
rdd_31_4	Memory Deserialized 1x Replicated	564.6 KiB	0.0 B	10.0.1.5:61782
rdd_31_5	Memory Deserialized 1x Replicated	564.3 KiB	0.0 B	10.0.1.5:61782
rdd_31_6	Memory Deserialized 1x Replicated	558.1 KiB	0.0 B	10.0.1.5:61782
rdd_31_7	Memory Deserialized 1x Replicated	557.6 KiB	0.0 B	10.0.1.5:61782

Figure 7-15. Spark UI showing cached table distribution across executor memory

SQL

The effects of Spark SQL queries that are executed as part of your Spark application are traceable and viewable through the SQL tab. You can see when the queries were executed and by which jobs, and their duration. For example, in our SortMergeJoin example we executed some queries; all of them are displayed in Figure 7-16, with links to drill further down.

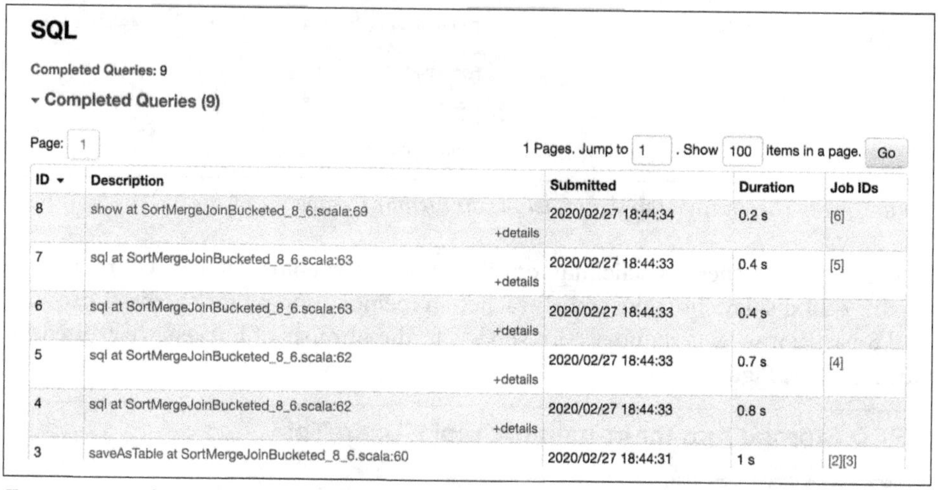

SQL

Completed Queries: 9

▾ Completed Queries (9)

Page: 1   1 Pages. Jump to 1 . Show 100 items in a page. Go

ID ▾	Description	Submitted	Duration	Job IDs
8	show at SortMergeJoinBucketed_8_6.scala:69 +details	2020/02/27 18:44:34	0.2 s	[6]
7	sql at SortMergeJoinBucketed_8_6.scala:63 +details	2020/02/27 18:44:33	0.4 s	[5]
6	sql at SortMergeJoinBucketed_8_6.scala:63 +details	2020/02/27 18:44:33	0.4 s	
5	sql at SortMergeJoinBucketed_8_6.scala:62 +details	2020/02/27 18:44:33	0.7 s	[4]
4	sql at SortMergeJoinBucketed_8_6.scala:62 +details	2020/02/27 18:44:33	0.8 s	
3	saveAsTable at SortMergeJoinBucketed_8_6.scala:60	2020/02/27 18:44:31	1 s	[2][3]

Figure 7-16. The SQL tab shows details on the completed SQL queries

Clicking on the description of a query displays details of the execution plan with all the physical operators, as shown in Figure 7-17. Under each physical operator of the plan—here, Scan In-memory table, HashAggregate, and Exchange—are SQL metrics.

These metrics are useful when we want to inspect the details of a physical operator and discover what transpired: how many rows were scanned, how many shuffle bytes were written, etc.

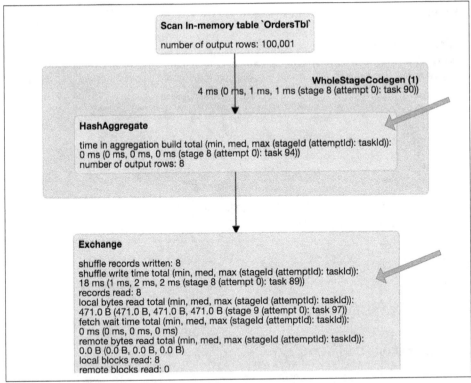

Figure 7-17. Spark UI showing detailed statistics on a SQL query

Environment

The Environment tab, shown in Figure 7-18, is just as important as the others. Knowing about the environment in which your Spark application is running reveals many clues that are useful for troubleshooting. In fact, it's imperative to know what environment variables are set, what jars are included, what Spark properties are set (and their respective values, especially if you tweaked some of the configs mentioned in "Optimizing and Tuning Spark for Efficiency" on page 173), what system properties are set, what runtime environment (such as JVM or Java version) is used, etc. All these read-only details are a gold mine of information supplementing your investigative efforts should you notice any abnormal behavior in your Spark application.

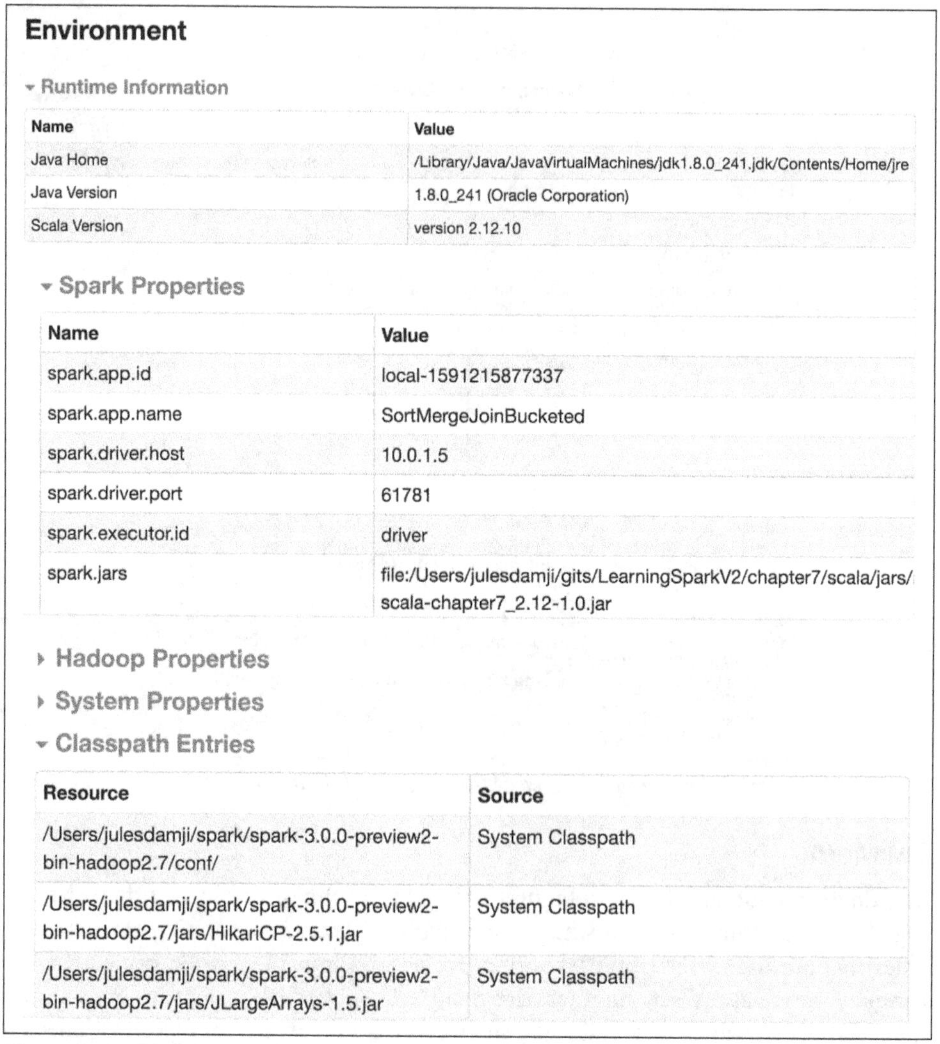

Environment

▼ Runtime Information

Name	Value
Java Home	/Library/Java/JavaVirtualMachines/jdk1.8.0_241.jdk/Contents/Home/jre
Java Version	1.8.0_241 (Oracle Corporation)
Scala Version	version 2.12.10

▼ Spark Properties

Name	Value
spark.app.id	local-1591215877337
spark.app.name	SortMergeJoinBucketed
spark.driver.host	10.0.1.5
spark.driver.port	61781
spark.executor.id	driver
spark.jars	file:/Users/julesdamji/gits/LearningSparkV2/chapter7/scala/jars/scala-chapter7_2.12-1.0.jar

▶ Hadoop Properties
▶ System Properties
▼ Classpath Entries

Resource	Source
/Users/julesdamji/spark/spark-3.0.0-preview2-bin-hadoop2.7/conf/	System Classpath
/Users/julesdamji/spark/spark-3.0.0-preview2-bin-hadoop2.7/jars/HikariCP-2.5.1.jar	System Classpath
/Users/julesdamji/spark/spark-3.0.0-preview2-bin-hadoop2.7/jars/JLargeArrays-1.5.jar	System Classpath

Figure 7-18. The Environment tab shows the runtime properties of your Spark cluster

Debugging Spark applications

In this section, we have navigated through the various tabs in the Spark UI. As you've seen, the UI provides a wealth of information that you can use for debugging and troubleshooting issues with your Spark applications. In addition to what we've covered here, it also provides access to both driver and executor stdout/stderr logs, where you might have logged debugging information.

Debugging through the UI is a different process than stepping through an application in your favorite IDE—more like sleuthing, following trails of bread crumbs—though if you prefer that approach, you can also debug a Spark application in an IDE such as IntelliJ IDEA (*https://oreil.ly/HkbIv*) on a local host.

The Spark 3.0 UI tabs (*https://oreil.ly/3X46q*) reveal insightful bread crumbs about what happened, along with access to both driver and executor stdout/stderr logs, where you might have logged debugging information.

Initially, this plethora of information can be overwhelming to a novice. But with time you'll gain an understanding of what to look for in each tab, and you'll begin to be able to detect and diagnose anomalies more quickly. Patterns will become clear, and by frequently visiting these tabs and getting familiar with them after running some Spark examples, you'll get accustomed to tuning and inspecting your Spark applications via the UI.

Summary

In this chapter we have discussed a number of optimization techniques for tuning your Spark applications. As you saw, by adjusting some of the default Spark configurations, you can improve scaling for large workloads, enhance parallelism, and minimize memory starvation among Spark executors. You also got a glimpse of how you can use caching and persisting strategies with appropriate levels to expedite access to your frequently used data sets, and we examined two commonly used joins Spark employs during complex aggregations and demonstrated how by bucketing DataFrames by sorted keys, you can skip over expensive shuffle operations.

Finally, to get a visual perspective on performance, the Spark UI completed the picture. Informative and detailed though the UI is, it's not equivalent to step-debugging in an IDE; yet we showed how you can become a Spark sleuth by examining and gleaning insights from the metrics and statistics, compute and memory usage data, and SQL query execution traces available on the half-dozen Spark UI tabs.

In the next chapter, we'll dive into Structured Streaming and show you how the Structured APIs that you learned about in earlier chapters allow you to write both streaming and batch applications in a continuous manner, enabling you to build reliable data lakes and pipelines.

Structured Streaming

In earlier chapters, you learned how to use structured APIs to process very large but finite volumes of data. However, often data arrives continuously and needs to be processed in a real-time manner. In this chapter, we will discuss how the same Structured APIs can be used for processing data streams as well.

Evolution of the Apache Spark Stream Processing Engine

Stream processing is defined as the continuous processing of endless streams of data. With the advent of big data, stream processing systems transitioned from single-node processing engines to multiple-node, distributed processing engines. Traditionally, distributed stream processing has been implemented with a *record-at-a-time processing model*, as illustrated in Figure 8-1.

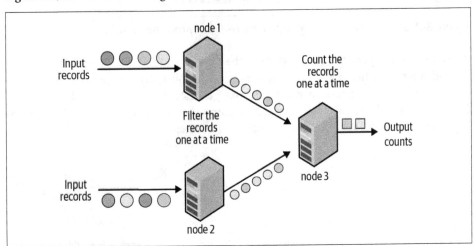

Figure 8-1. Traditional record-at-a-time processing model

The processing pipeline is composed of a directed graph of nodes, as shown in Figure 8-1; each node continuously receives one record at a time, processes it, and then forwards the generated record(s) to the next node in the graph. This processing model can achieve very low latencies—that is, an input record can be processed by the pipeline and the resulting output can be generated within milliseconds. However, this model is not very efficient at recovering from node failures and straggler nodes (i.e., nodes that are slower than others); it can either recover from a failure very fast with a lot of extra failover resources, or use minimal extra resources but recover slowly.[1]

The Advent of Micro-Batch Stream Processing

This traditional approach was challenged by Apache Spark when it introduced Spark Streaming (also called DStreams). It introduced the idea of *micro-batch stream processing*, where the streaming computation is modeled as a continuous series of small, map/reduce-style batch processing jobs (hence, "micro-batches") on small chunks of the stream data. This is illustrated in Figure 8-2.

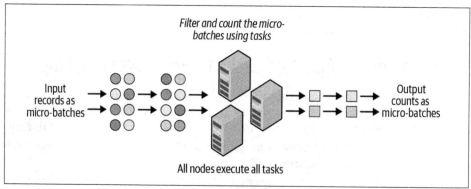

Figure 8-2. Structured Streaming uses a micro-batch processing model

As shown here, Spark Streaming divides the data from the input stream into, say, 1-second micro-batches. Each batch is processed in the Spark cluster in a distributed manner with small deterministic tasks that generate the output in micro-batches. Breaking down the streaming computation into these small tasks gives us two advantages over the traditional, continuous-operator model:

---

1 For a more detailed explanation, see the original research paper "Discretized Streams: Fault-Tolerant Streaming Computation at Scale" (*https://oreil.ly/Lz8mM*) by Matei Zaharia et al. (2013).

- Spark's agile task scheduling can very quickly and efficiently recover from failures and straggler executors by rescheduling one or more copies of the tasks on any of the other executors.

- The deterministic nature of the tasks ensures that the output data is the same no matter how many times the task is reexecuted. This crucial characteristic enables Spark Streaming to provide end-to-end exactly-once processing guarantees, that is, the generated output results will be such that every input record was processed exactly once.

This efficient fault tolerance does come at the cost of latency—the micro-batch model cannot achieve millisecond-level latencies; it usually achieves latencies of a few seconds (as low as half a second in some cases). However, we have observed that for an overwhelming majority of stream processing use cases, the benefits of micro-batch processing outweigh the drawback of second-scale latencies. This is because most streaming pipelines have at least one of the following characteristics:

- The pipeline does not need latencies lower than a few seconds. For example, when the streaming output is only going to be read by hourly jobs, it is not useful to generate output with subsecond latencies.

- There are larger delays in other parts of the pipeline. For example, if the writes by a sensor into Apache Kafka (a system for ingesting data streams) are batched to achieve higher throughput, then no amount of optimization in the downstream processing systems can make the end-to-end latency lower than the batching delays.

Furthermore, the DStream API was built upon Spark's batch RDD API. Therefore, DStreams had the same functional semantics and fault-tolerance model as RDDs. Spark Streaming thus proved that it is possible for a single, unified processing engine to provide consistent APIs and semantics for batch, interactive, and streaming workloads. This fundamental paradigm shift in stream processing propelled Spark Streaming to become one of the most widely used open source stream processing engines.

Lessons Learned from Spark Streaming (DStreams)

Despite all the advantages, the DStream API was not without its flaws. Here are a few key areas for improvement that were identified:

Lack of a single API for batch and stream processing
Even though DStreams and RDDs have consistent APIs (i.e., same operations and same semantics), developers still had to explicitly rewrite their code to use different classes when converting their batch jobs to streaming jobs.

Lack of separation between logical and physical plans
Spark Streaming executes the DStream operations in the same sequence in which they were specified by the developer. Since developers effectively specify the exact physical plan, there is no scope for automatic optimizations, and developers have to hand-optimize their code to get the best performance.

Lack of native support for event-time windows
DStreams define window operations based only on the time when each record is received by Spark Streaming (known as *processing time*). However, many use cases need to calculate windowed aggregates based on the time when the records were generated (known as *event time*) instead of when they were received or processed. The lack of native support of event-time windows made it hard for developers to build such pipelines with Spark Streaming.

These drawbacks shaped the design philosophy of Structured Streaming, which we will discuss next.

The Philosophy of Structured Streaming

Based on these lessons from DStreams, Structured Streaming was designed from scratch with one core philosophy—for developers, writing stream processing pipelines should be as easy as writing batch pipelines. In a nutshell, the guiding principles of Structured Streaming are:

A single, unified programming model and interface for batch and stream processing
This unified model offers a simple API interface for both batch and streaming workloads. You can use familiar SQL or batch-like DataFrame queries (like those you've learned about in the previous chapters) on your stream as you would on a batch, leaving dealing with the underlying complexities of fault tolerance, optimizations, and tardy data to the engine. In the coming sections, we will examine some of the queries you might write.

A broader definition of stream processing
Big data processing applications have grown complex enough that the line between real-time processing and batch processing has blurred significantly. The aim with Structured Streaming was to broaden its applicability from traditional stream processing to a larger class of applications; any application that periodically (e.g., every few hours) to continuously (like traditional streaming applications) processes data should be expressible using Structured Streaming.

Next, we'll discuss the programming model used by Structured Streaming.

The Programming Model of Structured Streaming

"Table" is a well-known concept that developers are familiar with when building batch applications. Structured Streaming extends this concept to streaming applications by treating a stream as an unbounded, continuously appended table, as illustrated in Figure 8-3.

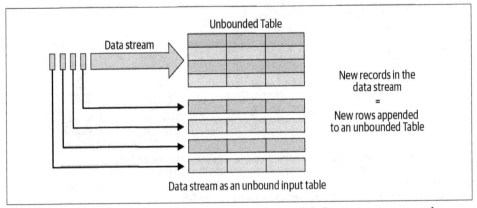

Figure 8-3. The Structured Streaming programming model: data stream as an unbounded table

Every new record received in the data stream is like a new row being appended to the unbounded input table. Structured Streaming will not actually retain all the input, but the output produced by Structured Streaming until time T will be equivalent to having all of the input until T in a static, bounded table and running a batch job on the table.

As shown in Figure 8-4, the developer then defines a query on this conceptual input table, as if it were a static table, to compute the result table that will be written to an output sink. Structured Streaming will automatically convert this batch-like query to a streaming execution plan. This is called *incrementalization*: Structured Streaming figures out what state needs to be maintained to update the result each time a record arrives. Finally, developers specify triggering policies to control when to update the results. Each time a trigger fires, Structured Streaming checks for new data (i.e., a new row in the input table) and incrementally updates the result.

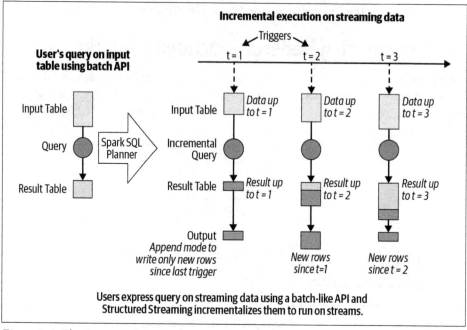

Figure 8-4. The Structured Streaming processing model

The last part of the model is the output mode. Each time the result table is updated, the developer will want to write the updates to an external system, such as a filesystem (e.g., HDFS, Amazon S3) or a database (e.g., MySQL, Cassandra). We usually want to write output incrementally. For this purpose, Structured Streaming provides three output modes:

Append mode

Only the new rows appended to the result table since the last trigger will be written to the external storage. This is applicable only in queries where existing rows in the result table cannot change (e.g., a map on an input stream).

Update mode

Only the rows that were updated in the result table since the last trigger will be changed in the external storage. This mode works for output sinks that can be updated in place, such as a MySQL table.

Complete mode

The entire updated result table will be written to external storage.

Unless complete mode is specified, the result table will not be fully materialized by Structured Streaming. Just enough information (known as "state") will be maintained to ensure that the changes in the result table can be computed and the updates can be output.

Thinking of the data streams as tables not only makes it easier to conceptualize the logical computations on the data, but also makes it easier to express them in code. Since Spark's DataFrame is a programmatic representation of a table, you can use the DataFrame API to express your computations on streaming data. All you need to do is define an input DataFrame (i.e., the input table) from a streaming data source, and then you apply operations on the DataFrame in the same way as you would on a DataFrame defined on a batch source.

In the next section, you will see how easy it is to write Structured Streaming queries using DataFrames.

The Fundamentals of a Structured Streaming Query

In this section, we are going to cover some high-level concepts that you'll need to understand to develop Structured Streaming queries. We will first walk through the key steps to define and start a streaming query, then we will discuss how to monitor the active query and manage its life cycle.

Five Steps to Define a Streaming Query

As discussed in the previous section, Structured Streaming uses the same DataFrame API as batch queries to express the data processing logic. However, there are a few key differences you need to know about for defining a Structured Streaming query. In this section, we will explore the steps involved in defining a streaming query by building a simple query that reads streams of text data over a socket and counts the words.

Step 1: Define input sources

As with batch queries, the first step is to define a DataFrame from a streaming source. However, when reading batch data sources, we need `spark.read` to create a `DataFrameReader`, whereas with streaming sources we need `spark.readStream` to create a `DataStreamReader`. `DataStreamReader` has most of the same methods as `DataFrameReader`, so you can use it in a similar way. Here is an example of creating a DataFrame from a text data stream to be received over a socket connection:

```python
# In Python
spark = SparkSession...
lines = (spark
  .readStream.format("socket")
  .option("host", "localhost")
  .option("port", 9999)
  .load())
```

```scala
// In Scala
val spark = SparkSession...
val lines = spark
  .readStream.format("socket")
  .option("host", "localhost")
  .option("port", 9999)
  .load()
```

This code generates the `lines` DataFrame as an unbounded table of newline-separated text data read from localhost:9999. Note that, similar to batch sources with `spark.read`, this does not immediately start reading the streaming data; it only sets up the configurations necessary for reading the data once the streaming query is explicitly started.

Besides sockets, Apache Spark natively supports reading data streams from Apache Kafka and all the various file-based formats that `DataFrameReader` supports (Parquet, ORC, JSON, etc.). The details of these sources and their supported options are discussed later in this chapter. Furthermore, a streaming query can define multiple input sources, both streaming and batch, which can be combined using DataFrame operations like unions and joins (also discussed later in this chapter).

Step 2: Transform data

Now we can apply the usual DataFrame operations, such as splitting the lines into individual words and then counting them, as shown in the following code:

```python
# In Python
from pyspark.sql.functions import *
words = lines.select(explode(split(col("value"), "\\s")).alias("word"))
counts = words.groupBy("word").count()
```

```scala
// In Scala
import org.apache.spark.sql.functions._
val words = lines.select(explode(split(col("value"), "\\s")).as("word"))
val counts = words.groupBy("word").count()
```

`counts` is a *streaming DataFrame* (that is, a DataFrame on unbounded, streaming data) that represents the running word counts that will be computed once the streaming query is started and the streaming input data is being continuously processed.

Note that these operations to transform the `lines` streaming DataFrame would work in the exact same way if `lines` were a batch DataFrame. In general, most DataFrame

---

operations that can be applied on a batch DataFrame can also be applied on a streaming DataFrame. To understand which operations are supported in Structured Streaming, you have to recognize the two broad classes of data transformations:

Stateless transformations

Operations like `select()`, `filter()`, `map()`, etc. do not require any information from previous rows to process the next row; each row can be processed by itself. The lack of previous "state" in these operations make them stateless. Stateless operations can be applied to both batch and streaming DataFrames.

Stateful transformations

In contrast, an aggregation operation like `count()` requires maintaining state to combine data across multiple rows. More specifically, any DataFrame operations involving grouping, joining, or aggregating are stateful transformations. While many of these operations are supported in Structured Streaming, a few combinations of them are not supported because it is either computationally hard or infeasible to compute them in an incremental manner.

The stateful operations supported by Structured Streaming and how to manage their state at runtime are discussed later in the chapter.

Step 3: Define output sink and output mode

After transforming the data, we can define how to write the processed output data with `DataFrame.writeStream` (instead of `DataFrame.write`, used for batch data). This creates a `DataStreamWriter` which, similar to `DataFrameWriter`, has additional methods to specify the following:

- Output writing details (where and how to write the output)
- Processing details (how to process data and how to recover from failures)

Let's start with the output writing details (we will focus on the processing details in the next step). For example, the following snippet shows how to write the final counts to the console:

```python
# In Python
writer = counts.writeStream.format("console").outputMode("complete")
```

```scala
// In Scala
val writer = counts.writeStream.format("console").outputMode("complete")
```

Here we have specified `"console"` as the output streaming sink and `"complete"` as the output mode. The output mode of a streaming query specifies what part of the updated output to write out after processing new input data. In this example, as a chunk of new input data is processed and the word counts are updated, we can choose to print to the console either the counts of all the words seen until now (that is, *complete mode*), or only those words that were updated in the last chunk of input data. This is decided by the specified output mode, which can be one of the following (as we already saw in "The Programming Model of Structured Streaming" on page 211:

Append mode

This is the default mode, where only the new rows added to the result table/Data-Frame (for example, the `counts` table) since the last trigger will be output to the sink. Semantically, this mode guarantees that any row that is output is never going to be changed or updated by the query in the future. Hence, append mode is supported by only those queries (e.g., stateless queries) that will never modify previously output data. In contrast, our word count query can update previously generated counts; therefore, it does not support append mode.

Complete mode

In this mode, all the rows of the result table/DataFrame will be output at the end of every trigger. This is supported by queries where the result table is likely to be much smaller than the input data and therefore can feasibly be retained in memory. For example, our word count query supports complete mode because the counts data is likely to be far smaller than the input data.

Update mode

In this mode, only the rows of the result table/DataFrame that were updated since the last trigger will be output at the end of every trigger. This is in contrast to append mode, as the output rows may be modified by the query and output again in the future. Most queries support update mode.

 Complete details on the output modes supported by different queries can be found in the latest Structured Streaming Programming Guide (*https://oreil.ly/hyuKL*).

Besides writing the output to the console, Structured Streaming natively supports streaming writes to files and Apache Kafka. In addition, you can write to arbitrary locations using the `foreachBatch()` and `foreach()` API methods. In fact, you can use `foreachBatch()` to write streaming outputs using existing batch data sources (but you will lose exactly-once guarantees). The details of these sinks and their supported options are discussed later in this chapter.

Step 4: Specify processing details

The final step before starting the query is to specify details of how to process the data. Continuing with our word count example, we are going to specify the processing details as follows:

```
# In Python
checkpointDir = "..."
writer2 = (writer
  .trigger(processingTime="1 second")
  .option("checkpointLocation", checkpointDir))

// In Scala
import org.apache.spark.sql.streaming._
val checkpointDir = "..."
val writer2 = writer
  .trigger(Trigger.ProcessingTime("1 second"))
  .option("checkpointLocation", checkpointDir)
```

Here we have specified two types of details using the `DataStreamWriter` that we created with `DataFrame.writeStream`:

Triggering details

This indicates when to trigger the discovery and processing of newly available streaming data. There are four options:

Default

When the trigger is not explicitly specified, then by default, the streaming query executes data in micro-batches where the next micro-batch is triggered as soon as the previous micro-batch has completed.

Processing time with trigger interval

You can explicitly specify the `ProcessingTime` trigger with an interval, and the query will trigger micro-batches at that fixed interval.

Once

In this mode, the streaming query will execute exactly one micro-batch—it processes all the new data available in a single batch and then stops itself. This is useful when you want to control the triggering and processing from an external scheduler that will restart the query using any custom schedule (e.g., to control cost by only executing a query once per day (*https://oreil.ly/ Y7EZy*)).

Continuous

This is an experimental mode (as of Spark 3.0) where the streaming query will process data continuously instead of in micro-batches. While only a small subset of DataFrame operations allow this mode to be used, it can provide much lower latency (as low as milliseconds) than the micro-batch trigger modes. Refer to the latest Structured Streaming Programming Guide (*https://oreil.ly/7cERT*) for the most up-to-date information.

Checkpoint location

This is a directory in any HDFS-compatible filesystem where a streaming query saves its progress information—that is, what data has been successfully processed. Upon failure, this metadata is used to restart the failed query exactly where it left off. Therefore, setting this option is necessary for failure recovery with exactly-once guarantees.

Step 5: Start the query

Once everything has been specified, the final step is to start the query, which you can do with the following:

```python
# In Python
streamingQuery = writer2.start()
```

```scala
// In Scala
val streamingQuery = writer2.start()
```

The returned object of type `streamingQuery` represents an active query and can be used to manage the query, which we will cover later in this chapter.

Note that `start()` is a nonblocking method, so it will return as soon as the query has started in the background. If you want the main thread to block until the streaming query has terminated, you can use `streamingQuery.awaitTermination()`. If the query fails in the background with an error, `awaitTermination()` will also fail with that same exception.

You can wait up to a timeout duration using `awaitTermination(timeoutMillis)`, and you can explicitly stop the query with `streamingQuery.stop()`.

Putting it all together

To summarize, here is the complete code for reading streams of text data over a socket, counting the words, and printing the counts to the console:

```python
# In Python
from pyspark.sql.functions import *
spark = SparkSession...
lines = (spark
  .readStream.format("socket")
  .option("host", "localhost")
  .option("port", 9999)
  .load())

words = lines.select(explode(split(col("value"), "\\s")).alias("word"))
counts = words.groupBy("word").count()
checkpointDir = "..."
streamingQuery = (counts
  .writeStream
  .format("console")
```

```
    .outputMode("complete")
    .trigger(processingTime="1 second")
    .option("checkpointLocation", checkpointDir)
    .start())
streamingQuery.awaitTermination()

// In Scala
import org.apache.spark.sql.functions._
import org.apache.spark.sql.streaming._
val spark = SparkSession...
val lines = spark
    .readStream.format("socket")
    .option("host", "localhost")
    .option("port", 9999)
    .load()

val words = lines.select(explode(split(col("value"), "\\s")).as("word"))
val counts = words.groupBy("word").count()

val checkpointDir = "..."
val streamingQuery = counts.writeStream
    .format("console")
    .outputMode("complete")
    .trigger(Trigger.ProcessingTime("1 second"))
    .option("checkpointLocation", checkpointDir)
    .start()
streamingQuery.awaitTermination()
```

After the query has started, a background thread continuously reads new data from the streaming source, processes it, and writes it to the streaming sinks. Next, let's take a quick peek under the hood at how this is executed.

Under the Hood of an Active Streaming Query

Once the query starts, the following sequence of steps transpires in the engine, as depicted in Figure 8-5. The DataFrame operations are converted into a logical plan, which is an abstract representation of the computation that Spark SQL uses to plan a query:

1. Spark SQL analyzes and optimizes this logical plan to ensure that it can be executed incrementally and efficiently on streaming data.

2. Spark SQL starts a background thread that continuously executes the following loop:[2]

---

2 This execution loop runs for micro-batch-based trigger modes (i.e., ProcessingTime and Once), but not for the Continuous trigger mode.

a. Based on the configured trigger interval, the thread checks the streaming sources for the availability of new data.

b. If available, the new data is executed by running a micro-batch. From the optimized logical plan, an optimized Spark execution plan is generated that reads the new data from the source, incrementally computes the updated result, and writes the output to the sink according to the configured output mode.

c. For every micro-batch, the exact range of data processed (e.g., the set of files or the range of Apache Kafka offsets) and any associated state are saved in the configured checkpoint location so that the query can deterministically reprocess the exact range if needed.

3. This loop continues until the query is terminated, which can occur for one of the following reasons:

a. A failure has occurred in the query (either a processing error or a failure in the cluster).

b. The query is explicitly stopped using `streamingQuery.stop()`.

c. If the trigger is set to Once, then the query will stop on its own after executing a single micro-batch containing all the available data.

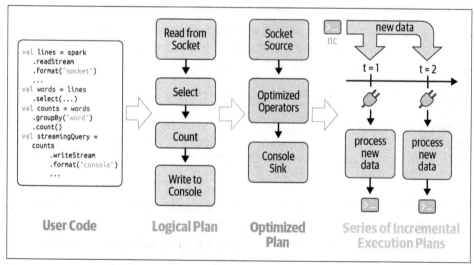

Figure 8-5. Incremental execution of streaming queries

A key point you should remember about Structured Streaming is that underneath it is using Spark SQL to execute the data. As such, the full power of Spark SQL's hyperoptimized execution engine is utilized to maximize the stream processing throughput, providing key performance advantages.

Next, we will discuss how to restart a streaming query after termination and the life cycle of a streaming query.

Recovering from Failures with Exactly-Once Guarantees

To restart a terminated query in a completely new process, you have to create a new SparkSession, redefine all the DataFrames, and start the streaming query on the final result using the same checkpoint location as the one used when the query was started the first time. For our word count example, you can simply reexecute the entire code snippet shown earlier, from the definition of spark in the first line to the final start() in the last line.

The checkpoint location must be the same across restarts because this directory contains the unique identity of a streaming query and determines the life cycle of the query. If the checkpoint directory is deleted or the same query is started with a different checkpoint directory, it is like starting a new query from scratch. Specifically, checkpoints have record-level information (e.g., Apache Kafka offsets) to track the data range the last incomplete micro-batch was processing. The restarted query will use this information to start processing records precisely after the last successfully completed micro-batch. If the previous query had planned a micro-batch but had terminated before completion, then the restarted query will reprocess the same range of data before processing new data. Coupled with Spark's deterministic task execution, the regenerated output will be the same as it was expected to be before the restart.

Structured Streaming can ensure *end-to-end exactly-once guarantees* (that is, the output is as if each input record was processed exactly once) when the following conditions have been satisfied:

Replayable streaming sources
   The data range of the last incomplete micro-batch can be reread from the source.

Deterministic computations
   All data transformations deterministically produce the same result when given the same input data.

Idempotent streaming sink
   The sink can identify reexecuted micro-batches and ignore duplicate writes that may be caused by restarts.

Note that our word count example does not provide exactly-once guarantees because the socket source is not replayable and the console sink is not idempotent.

As a final note regarding restarting queries, it is possible to make minor modifications to a query between restarts. Here are a few ways you can modify the query:

DataFrame transformations

You can make minor modifications to the transformations between restarts. For example, in our streaming word count example, if you want to ignore lines that have corrupted byte sequences that can crash the query, you can add a filter in the transformation:

```
# In Python
# isCorruptedUdf = udf to detect corruption in string

filteredLines = lines.filter("isCorruptedUdf(value) = false")
words = lines.select(explode(split(col("value"), "\\s").alias("word")))

// In Scala
// val isCorruptedUdf = udf to detect corruption in string

val filteredLines = lines.filter("isCorruptedUdf(value) = false")
val words = lines.select(explode(split(col("value"), "\\s").as("word")))
```

Upon restarting with this modified words DataFrame, the restarted query will apply the filter on all data processed since the restart (including the last incomplete micro-batch), preventing the query from failing again.

Source and sink options

Whether a `readStream` or `writeStream` option can be changed between restarts depends on the semantics of the specific source or sink. For example, you should not change the `host` and `port` options for the socket source if data is going to be sent to that host and port. But you can add an option to the console sink to print up to one hundred changed counts after every trigger:

```
writeStream.format("console").option("numRows", "100")...
```

Processing details

As discussed earlier, the checkpoint location must not be changed between restarts. However, other details like trigger interval can be changed without breaking fault-tolerance guarantees.

For more information on the narrow set of changes that are allowed between restarts, see the latest Structured Streaming Programming Guide (*https://oreil.ly/am885*).

Monitoring an Active Query

An important part of running a streaming pipeline in production is tracking its health. Structured Streaming provides several ways to track the status and processing metrics of an active query.

Querying current status using StreamingQuery

You can query the current health of an active query using the `StreamingQuery` instance. Here are two methods:

Get current metrics using StreamingQuery. When a query processes some data in a micro-batch, we consider it to have made some progress. `lastProgress()` returns information on the last completed micro-batch. For example, printing the returned object (`StreamingQueryProgress` in Scala/Java or a dictionary in Python) will produce something like this:

```
// In Scala/Python
{
  "id" : "ce011fdc-8762-4dcb-84eb-a77333e28109",
  "runId" : "88e2ff94-ede0-45a8-b687-6316fbef529a",
  "name" : "MyQuery",
  "timestamp" : "2016-12-14T18:45:24.873Z",
  "numInputRows" : 10,
  "inputRowsPerSecond" : 120.0,
  "processedRowsPerSecond" : 200.0,
  "durationMs" : {
    "triggerExecution" : 3,
    "getOffset" : 2
  },
  "stateOperators" : [ ],
  "sources" : [ {
    "description" : "KafkaSource[Subscribe[topic-0]]",
    "startOffset" : {
      "topic-0" : {
        "2" : 0,
        "1" : 1,
        "0" : 1
      }
    },
    "endOffset" : {
      "topic-0" : {
        "2" : 0,
        "1" : 134,
        "0" : 534
      }
    },
    "numInputRows" : 10,
    "inputRowsPerSecond" : 120.0,
    "processedRowsPerSecond" : 200.0
```

```
  } ],
  "sink" : {
    "description" : "MemorySink"
  }
}
```

Some of the noteworthy columns are:

id
>   Unique identifier tied to a checkpoint location. This stays the same throughout the lifetime of a query (i.e., across restarts).

runId
>   Unique identifier for the current (re)started instance of the query. This changes with every restart.

numInputRows
>   Number of input rows that were processed in the last micro-batch.

inputRowsPerSecond
>   Current rate at which input rows are being generated at the source (average over the last micro-batch duration).

processedRowsPerSecond
>   Current rate at which rows are being processed and written out by the sink (average over the last micro-batch duration). If this rate is consistently lower than the input rate, then the query is unable to process data as fast as it is being generated by the source. This is a key indicator of the health of the query.

sources *and* sink
>   Provides source/sink-specific details of the data processed in the last batch.

Get current status using StreamingQuery.status(). This provides information on what the background query thread is doing at this moment. For example, printing the returned object will produce something like this:

```
// In Scala/Python
{
  "message" : "Waiting for data to arrive",
  "isDataAvailable" : false,
  "isTriggerActive" : false
}
```

Publishing metrics using Dropwizard Metrics

Spark supports reporting metrics via a popular library called Dropwizard Metrics (*https://metrics.dropwizard.io*). This library allows metrics to be published to many popular monitoring frameworks (Ganglia, Graphite, etc.). These metrics are by default not enabled for Structured Streaming queries due to their high volume of reported data. To enable them, apart from configuring Dropwizard Metrics for Spark (*https://oreil.ly/4xenP*), you have to explicitly set the `SparkSession` configuration `spark.sql.streaming.metricsEnabled` to `true` before starting your query.

Note that only a subset of the information available through `StreamingQuery.lastProgress()` is published through Dropwizard Metrics. If you want to continuously publish more progress information to arbitrary locations, you have to write custom listeners, as discussed next.

Publishing metrics using custom StreamingQueryListeners

`StreamingQueryListener` is an event listener interface with which you can inject arbitrary logic to continuously publish metrics. This developer API is available only in Scala/Java. There are two steps to using custom listeners:

1. Define your custom listener. The `StreamingQueryListener` interface provides three methods that can be defined by your implementation to get three types of events related to a streaming query: start, progress (i.e., a trigger was executed), and termination. Here is an example:

   ```
   // In Scala
   import org.apache.spark.sql.streaming._
   val myListener = new StreamingQueryListener() {
     override def onQueryStarted(event: QueryStartedEvent): Unit = {
       println("Query started: " + event.id)
     }
     override def onQueryTerminated(event: QueryTerminatedEvent): Unit = {
       println("Query terminated: " + event.id)
     }
     override def onQueryProgress(event: QueryProgressEvent): Unit = {
       println("Query made progress: " + event.progress)
     }
   }
   ```

2. Add your listener to the `SparkSession` before starting the query:

   ```
   // In Scala
   spark.streams.addListener(myListener)
   ```

After adding the listener, all events of streaming queries running on this `Spark Session` will start calling the listener's methods.

Streaming Data Sources and Sinks

Now that we have covered the basic steps you need to express an end-to-end Structured Streaming query, let's examine how to use the built-in streaming data sources and sinks. As a reminder, you can create DataFrames from streaming sources using `SparkSession.readStream()` and write the output from a result DataFrame using `DataFrame.writeStream()`. In each case, you can specify the source type using the method `format()`. We will see a few concrete examples later.

Files

Structured Streaming supports reading and writing data streams to and from files in the same formats as the ones supported in batch processing: plain text, CSV, JSON, Parquet, ORC, etc. Here we will discuss how to operate Structured Streaming on files.

Reading from files

Structured Streaming can treat files written into a directory as a data stream. Here is an example:

```python
# In Python
from pyspark.sql.types import *
inputDirectoryOfJsonFiles =  ...

fileSchema = (StructType()
  .add(StructField("key", IntegerType()))
  .add(StructField("value", IntegerType())))

inputDF = (spark
  .readStream
  .format("json")
  .schema(fileSchema)
  .load(inputDirectoryOfJsonFiles))
```

```scala
// In Scala
import org.apache.spark.sql.types._
val inputDirectoryOfJsonFiles =  ...

val fileSchema = new StructType()
  .add("key", IntegerType)
  .add("value", IntegerType)

val inputDF = spark.readStream
  .format("json")
  .schema(fileSchema)
  .load(inputDirectoryOfJsonFiles)
```

The returned streaming DataFrame will have the specified schema. Here are a few key points to remember when using files:

- All the files must be of the same format and are expected to have the same schema. For example, if the format is `"json"`, all the files must be in the JSON format with one JSON record per line. The schema of each JSON record must match the one specified with `readStream()`. Violation of these assumptions can lead to incorrect parsing (e.g., unexpected `null` values) or query failures.

- Each file must appear in the directory listing atomically—that is, the whole file must be available at once for reading, and once it is available, the file cannot be updated or modified. This is because Structured Streaming will process the file when the engine finds it (using directory listing) and internally mark it as processed. Any changes to that file will not be processed.

- When there are multiple new files to process but it can only pick some of them in the next micro-batch (e.g., because of rate limits), it will select the files with the earliest timestamps. Within the micro-batch, however, there is no predefined order of reading of the selected files; all of them will be read in parallel.

> This streaming file source supports a number of common options, including the file format–specific options supported by `spark.read()` (see "Data Sources for DataFrames and SQL Tables" on page 94 in Chapter 4) and several streaming-specific options (e.g., `maxFilesPerTrigger` to limit the file processing rate). See the programming guide (*https://oreil.ly/VxU9U*) for full details.

Writing to files

Structured Streaming supports writing streaming query output to files in the same formats as reads. However, it only supports append mode, because while it is easy to write new files in the output directory (i.e., append data to a directory), it is hard to modify existing data files (as would be expected with update and complete modes). It also supports partitioning. Here is an example:

```
# In Python
outputDir = ...
checkpointDir = ...
resultDF = ...

streamingQuery = (resultDF.writeStream
  .format("parquet")
  .option("path", outputDir)
  .option("checkpointLocation", checkpointDir)
  .start())
```

```scala
// In Scala
val outputDir = ...
val checkpointDir = ...
val resultDF = ...

val streamingQuery = resultDF
  .writeStream
  .format("parquet")
  .option("path", outputDir)
  .option("checkpointLocation", checkpointDir)
  .start()
```

Instead of using the "path" option, you can specify the output directory directly as start(outputDir).

A few key points to remember:

- Structured Streaming achieves end-to-end exactly-once guarantees when writing to files by maintaining a log of the data files that have been written to the directory. This log is maintained in the subdirectory _spark_metadata. Any Spark query on the directory (not its subdirectories) will automatically use the log to read the correct set of data files so that the exactly-once guarantee is maintained (i.e., no duplicate data or partial files are read). Note that other processing engines may not be aware of this log and hence may not provide the same guarantee.

- If you change the schema of the result DataFrame between restarts, then the output directory will have data in multiple schemas. These schemas have to be reconciled when querying the directory.

Apache Kafka

Apache Kafka (*https://kafka.apache.org/*) is a popular publish/subscribe system that is widely used for storage of data streams. Structured Streaming has built-in support for reading from and writing to Apache Kafka.

Reading from Kafka

To perform distributed reads from Kafka, you have to use options to specify how to connect to the source. Say you want to subscribe to data from the topic "events". Here is how you can create a streaming DataFrame:

```python
# In Python
inputDF = (spark
  .readStream
  .format("kafka")
  .option("kafka.bootstrap.servers", "host1:port1,host2:port2")
  .option("subscribe", "events")
  .load())
```

```
// In Scala
val inputDF = spark
  .readStream
  .format("kafka")
  .option("kafka.bootstrap.servers", "host1:port1,host2:port2")
  .option("subscribe", "events")
  .load()
```

The returned DataFrame will have the schema described in Table 8-1.

Table 8-1. Schema of the DataFrame generated by the Kafka source

Column name	Column type	Description
key	binary	Key data of the record as bytes.
value	binary	Value data of the record as bytes.
topic	string	Kafka topic the record was in. This is useful when subscribed to multiple topics.
partition	int	Partition of the Kafka topic the record was in.
offset	long	Offset value of the record.
timestamp	long	Timestamp associated with the record.
timestampType	int	Enumeration for the type of the timestamp associated with the record.

You can also choose to subscribe to multiple topics, a pattern of topics, or even a specific partition of a topic. Furthermore, you can choose whether to read only new data in the subscribed-to topics or process all the available data in those topics. You can even read Kafka data from batch queries—that is, treat Kafka topics like tables. See the Kafka Integration Guide (*https://oreil.ly/FVP0l*) for more details.

Writing to Kafka

For writing to Kafka, Structured Streaming expects the result DataFrame to have a few columns of specific names and types, as outlined in Table 8-2.

Table 8-2. Schema of DataFrame that can be written to the Kafka sink

Column name	Column type	Description
key (optional)	string or binary	If present, the bytes will be written as the Kafka record key; otherwise, the key will be empty.
value (required)	string or binary	The bytes will be written as the Kafka record value.
topic (required only if "topic" is not specified as option)	string	If "topic" is not specified as an option, this determines the topic to write the key/value to. This is useful for fanning out the writes to multiple topics. If the "topic" option has been specified, this value is ignored.

You can write to Kafka in all three output modes, though complete mode is not recommended as it will repeatedly output the same records. Here is a concrete example of writing the output of our earlier word count query into Kafka in update mode:

```python
# In Python
counts = ... # DataFrame[word: string, count: long]
streamingQuery = (counts
  .selectExpr(
    "cast(word as string) as key",
    "cast(count as string) as value")
  .writeStream
  .format("kafka")
  .option("kafka.bootstrap.servers", "host1:port1,host2:port2")
  .option("topic", "wordCounts")
  .outputMode("update")
  .option("checkpointLocation", checkpointDir)
  .start())
```

```scala
// In Scala
val counts = ... // DataFrame[word: string, count: long]
val streamingQuery = counts
  .selectExpr(
    "cast(word as string) as key",
    "cast(count as string) as value")
  .writeStream
  .format("kafka")
  .option("kafka.bootstrap.servers", "host1:port1,host2:port2")
  .option("topic", "wordCounts")
  .outputMode("update")
  .option("checkpointLocation", checkpointDir)
  .start()
```

See the Kafka Integration Guide (*https://oreil.ly/tFo-N*) for more details.

Custom Streaming Sources and Sinks

In this section, we will discuss how to read and write to storage systems that do not have built-in support in Structured Streaming. In particular, you'll see how to use the foreachBatch() and foreach() methods to implement custom logic to write to your storage.

Writing to any storage system

There are two operations that allow you to write the output of a streaming query to arbitrary storage systems: foreachBatch() and foreach(). They have slightly different use cases: while foreach() allows custom write logic on every row, foreach Batch() allows arbitrary operations and custom logic on the output of each micro-batch. Let's explore their usage in more detail.

Using foreachBatch(). `foreachBatch()` allows you to specify a function that is executed on the output of every micro-batch of a streaming query. It takes two parameters: a DataFrame or Dataset that has the output of a micro-batch, and the unique identifier of the micro-batch. As an example, say we want to write the output of our earlier word count query to Apache Cassandra (*http://cassandra.apache.org/*). As of Spark Cassandra Connector 2.4.2 (*https://oreil.ly/I7Mof*), there is no support for writing streaming DataFames. But you can use the connector's batch DataFrame support to write the output of each batch (i.e., updated word counts) to Cassandra, as shown here:

```python
# In Python
hostAddr = "<ip address>"
keyspaceName = "<keyspace>"
tableName = "<tableName>"

spark.conf.set("spark.cassandra.connection.host", hostAddr)

def writeCountsToCassandra(updatedCountsDF, batchId):
    # Use Cassandra batch data source to write the updated counts
    (updatedCountsDF
      .write
      .format("org.apache.spark.sql.cassandra")
      .mode("append")
      .options(table=tableName, keyspace=keyspaceName)
      .save())

streamingQuery = (counts
  .writeStream
  .foreachBatch(writeCountsToCassandra)
  .outputMode("update")
  .option("checkpointLocation", checkpointDir)
  .start())
```

```scala
// In Scala
import org.apache.spark.sql.DataFrame

val hostAddr = "<ip address>"
val keyspaceName = "<keyspace>"
val tableName = "<tableName>"

spark.conf.set("spark.cassandra.connection.host", hostAddr)

def writeCountsToCassandra(updatedCountsDF: DataFrame, batchId: Long) {
    // Use Cassandra batch data source to write the updated counts
    updatedCountsDF
      .write
      .format("org.apache.spark.sql.cassandra")
      .options(Map("table" -> tableName, "keyspace" -> keyspaceName))
      .mode("append")
      .save()
```

```
      }

val streamingQuery = counts
  .writeStream
  .foreachBatch(writeCountsToCassandra _)
  .outputMode("update")
  .option("checkpointLocation", checkpointDir)
  .start()
```

With `foreachBatch()`, you can do the following:

Reuse existing batch data sources

As shown in the previous example, with `foreachBatch()` you can use existing batch data sources (i.e., sources that support writing batch DataFrames) to write the output of streaming queries.

Write to multiple locations

If you want to write the output of a streaming query to multiple locations (e.g., an OLAP data warehouse and an OLTP database), then you can simply write the output DataFrame/Dataset multiple times. However, each attempt to write can cause the output data to be recomputed (including possible rereading of the input data). To avoid recomputations, you should cache the `batchOutputData Frame`, write it to multiple locations, and then uncache it:

```
# In Python
def writeCountsToMultipleLocations(updatedCountsDF, batchId):
  updatedCountsDF.persist()
  updatedCountsDF.write.format(...).save()  # Location 1
  updatedCountsDF.write.format(...).save()  # Location 2
  updatedCountsDF.unpersist()

// In Scala
def writeCountsToMultipleLocations(
  updatedCountsDF: DataFrame,
  batchId: Long) {
    updatedCountsDF.persist()
    updatedCountsDF.write.format(...).save()  // Location 1
    updatedCountsDF.write.format(...).save()  // Location 2
    updatedCountsDF.unpersist()
  }
```

Apply additional DataFrame operations

Many DataFrame API operations are not supported[3] on streaming DataFrames because Structured Streaming does not support generating incremental plans in those cases. Using `foreachBatch()`, you can apply some of these operations on

---

3 For the full list of unsupported operations, see the Structured Streaming Programming Guide (*https://oreil.ly/ wa60L*).

each micro-batch output. However, you will have to reason about the end-to-end semantics of doing the operation yourself.

 foreachBatch() only provides at-least-once write guarantees. You can get exactly-once guarantees by using the batchId to deduplicate multiple writes from reexecuted micro-batches.

Using foreach(). If foreachBatch() is not an option (for example, if a corresponding batch data writer does not exist), then you can express your custom writer logic using foreach(). Specifically, you can express the data-writing logic by dividing it into three methods: open(), process(), and close(). Structured Streaming will use these methods to write each partition of the output records. Here is an abstract example:

```python
# In Python
# Variation 1: Using function
def process_row(row):
    # Write row to storage
    pass

query = streamingDF.writeStream.foreach(process_row).start()

# Variation 2: Using the ForeachWriter class
class ForeachWriter:
  def open(self, partitionId, epochId):
    # Open connection to data store
    # Return True if write should continue
    # This method is optional in Python
    # If not specified, the write will continue automatically
    return True

  def process(self, row):
    # Write string to data store using opened connection
    # This method is NOT optional in Python
    pass

  def close(self, error):
    # Close the connection. This method is optional in Python
    pass

resultDF.writeStream.foreach(ForeachWriter()).start()
```

```scala
// In Scala
import org.apache.spark.sql.ForeachWriter
val foreachWriter = new ForeachWriter[String] {  // typed with Strings

    def open(partitionId: Long, epochId: Long): Boolean = {
      // Open connection to data store
      // Return true if write should continue
```

```
      }

      def process(record: String): Unit = {
        // Write string to data store using opened connection
      }

      def close(errorOrNull: Throwable): Unit = {
        // Close the connection
      }
    }

  resultDSofStrings.writeStream.foreach(foreachWriter).start()
```

The detailed semantics of these methods as executed are discussed in the Structured Streaming Programming Guide (*https://oreil.ly/dL7mc*).

Reading from any storage system

Unfortunately, as of Spark 3.0, the APIs to build custom streaming sources and sinks are still experimental. The DataSourceV2 initiative in Spark 3.0 introduces the streaming APIs but they are yet to be declared as stable. Hence, there is no official way to read from arbitrary storage systems.

Data Transformations

In this section, we are going to dig deeper into the data transformations supported in Structured Streaming. As briefly discussed earlier, only the DataFrame operations that can be executed incrementally are supported in Structured Streaming. These operations are broadly classified into *stateless* and *stateful* operations. We will define each type of operation and explain how to identify which operations are stateful.

Incremental Execution and Streaming State

As we discussed in "Under the Hood of an Active Streaming Query" on page 219, the Catalyst optimizer in Spark SQL converts all the DataFrame operations to an optimized logical plan. The Spark SQL planner, which decides how to execute a logical plan, recognizes that this is a streaming logical plan that needs to operate on continuous data streams. Accordingly, instead of converting the logical plan to a one-time physical execution plan, the planner generates a continuous sequence of execution plans. Each execution plan updates the final result DataFrame incrementally—that is, the plan processes only a chunk of new data from the input streams and possibly some intermediate, partial result computed by the previous execution plan.

Each execution is considered as a micro-batch, and the partial intermediate result that is communicated between the executions is called the streaming "state." Data-Frame operations can be broadly classified into stateless and stateful operations based on whether executing the operation incrementally requires maintaining a state. In the

---

rest of this section, we are going to explore the distinction between stateless and stateful operations and how their presence in a streaming query requires different runtime configuration and resource management.

> Some logical operations are fundamentally either impractical or very expensive to compute incrementally, and hence they are not supported in Structured Streaming. For example, any attempt to start a streaming query with an operation like cube() or rollup() will throw an UnsupportedOperationException.

Stateless Transformations

All projection operations (e.g., select(), explode(), map(), flatMap()) and selection operations (e.g., filter(), where()) process each input record individually without needing any information from previous rows. This lack of dependence on prior input data makes them stateless operations.

A streaming query having only stateless operations supports the append and update output modes, but not complete mode. This makes sense: since any processed output row of such a query cannot be modified by any future data, it can be written out to all streaming sinks in append mode (including append-only ones, like files of any format). On the other hand, such queries naturally do not combine information across input records, and therefore may not reduce the volume of the data in the result. Complete mode is not supported because storing the ever-growing result data is usually costly. This is in sharp contrast with stateful transformations, as we will discuss next.

Stateful Transformations

The simplest example of a stateful transformation is DataFrame.groupBy().count(), which generates a running count of the number of records received since the beginning of the query. In every micro-batch, the incremental plan adds the count of new records to the previous count generated by the previous micro-batch. This partial count communicated between plans is the state. This state is maintained in the memory of the Spark executors and is checkpointed to the configured location in order to tolerate failures. While Spark SQL automatically manages the life cycle of this state to ensure correct results, you typically have to tweak a few knobs to control the resource usage for maintaining state. In this section, we are going to explore how different stateful operators manage their state under the hood.

Distributed and fault-tolerant state management

Recall from Chapters 1 and 2 that a Spark application running in a cluster has a driver and one or more executors. Spark's scheduler running in the driver breaks down your high-level operations into smaller tasks and puts them in task queues, and as resources become available, the executors pull the tasks from the queues to execute them. Each micro-batch in a streaming query essentially performs one such set of tasks that read new data from streaming sources and write updated output to streaming sinks. For stateful stream processing queries, besides writing to sinks, each micro-batch of tasks generates intermediate state data which will be consumed by the next micro-batch. This state data generation is completely partitioned and distributed (as all reading, writing, and processing is in Spark), and it is cached in the executor memory for efficient consumption. This is illustrated in Figure 8-6, which shows how the state is managed in our original streaming word count query.

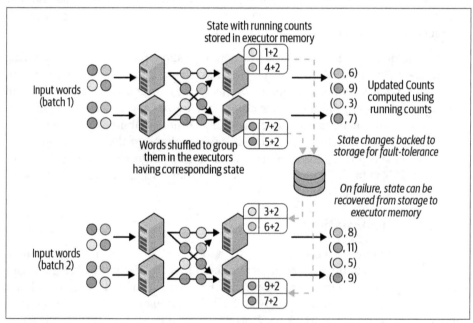

Figure 8-6. Distributed state management in Structured Streaming

Each micro-batch reads a new set of words, shuffles them within the executors to group them, computes the counts within the micro-batch, and finally adds them to the running counts to produce the new counts. These new counts are both the output and the state for the next micro-batch, and hence they are cached in the memory of the executors. The next micro-batch of data is grouped between executors in exactly the same way as before, so that each word is always processed by the same executor, and can therefore locally read and update its running count.

However, it is not sufficient to just keep this state in memory, as any failure (either of an executor or of the entire application) will cause the in-memory state to be lost. To avoid loss, we synchronously save the key/value state update as change logs in the checkpoint location provided by the user. These changes are co-versioned with the offset ranges processed in each batch, and the required version of the state can be automatically reconstructed by reading the checkpointed logs. In case of any failure, Structured Streaming is able to re-execute the failed micro-batch by reprocessing the same input data along with the same state that it had before that micro-batch, thus producing the same output data as it would have if there had been no failure. This is critical for ensuring end-to-end exactly-once guarantees.

To summarize, for all stateful operations, Structured Streaming ensures the correctness of the operation by automatically saving and restoring the state in a distributed manner. Depending on the stateful operation, all you may have to do is tune the state cleanup policy such that old keys and values can be automatically dropped from the cached state. This is what we will discuss next.

Types of stateful operations

The essence of streaming state is to retain summaries of past data. Sometimes old summaries need to be cleaned up from the state to make room for new summaries. Based on how this is done, we can distinguish two types of stateful operations:

Managed stateful operations
These automatically identify and clean up old state, based on an operation-specific definition of "old." You can tune what is defined as old in order to control the resource usage (e.g., executor memory used to store state). The operations that fall into this category are those for:

- Streaming aggregations
- Stream–stream joins
- Streaming deduplication

Unmanaged stateful operations
These operations let you define your own custom state cleanup logic. The operations in this category are:

- `MapGroupsWithState`
- `FlatMapGroupsWithState`

These operations allow you to define arbitrary stateful operations (sessionization, etc.).

Each of these operations are discussed in detail in the following sections.

Stateful Streaming Aggregations

Structured Streaming can incrementally execute most DataFrame aggregation operations. You can aggregate data by keys (e.g., streaming word count) and/or by time
(e.g., count records received every hour). In this section, we are going to discuss the
semantics and operational details of tuning these different types of streaming aggregations. We'll also briefly discuss the few types of aggregations that are not supported
in streaming. Let's begin with aggregations not involving time.

Aggregations Not Based on Time

Aggregations not involving time can be broadly classified into two categories:

Global aggregations

Aggregations across all the data in the stream. For example, say you have a
stream of sensor readings as a streaming DataFrame named sensorReadings.
You can calculate the running count of the total number of readings received
with the following query:

```
# In Python
runningCount = sensorReadings.groupBy().count()
```

```
// In Scala
val runningCount = sensorReadings.groupBy().count()
```

> You cannot use direct aggregation operations like Data
> Frame.count() and Dataset.reduce() on streaming Data-
> Frames. This is because, for static DataFrames, these
> operations immediately return the final computed aggregates,
> whereas for streaming DataFrames the aggregates have to be
> continuously updated. Therefore, you have to always use Data
> Frame.groupBy() or Dataset.groupByKey() for aggregations
> on streaming DataFrames.

Grouped aggregations

Aggregations within each group or key present in the data stream. For example, if
sensorReadings contains data from multiple sensors, you can calculate the running average reading of each sensor (say, for setting up a baseline value for each
sensor) with the following:

```
# In Python
baselineValues = sensorReadings.groupBy("sensorId").mean("value")
```

```
// In Scala
val baselineValues = sensorReadings.groupBy("sensorId").mean("value")
```

Besides counts and averages, streaming DataFrames support the following types of aggregations (similar to batch DataFrames):

All built-in aggregation functions
> `sum()`, `mean()`, `stddev()`, `countDistinct()`, `collect_set()`, `approx_count_dis tinct()`, etc. Refer to the API documentation (Python (*https://oreil.ly/olWT0*) and Scala (*https://oreil.ly/gvoeK*)) for more details.

Multiple aggregations computed together
> You can apply multiple aggregation functions to be computed together in the following manner:

```python
# In Python
from pyspark.sql.functions import *
multipleAggs = (sensorReadings
  .groupBy("sensorId")
  .agg(count("*"), mean("value").alias("baselineValue"),
    collect_set("errorCode").alias("allErrorCodes")))
```

```scala
// In Scala
import org.apache.spark.sql.functions.*
val multipleAggs = sensorReadings
  .groupBy("sensorId")
  .agg(count("*"), mean("value").alias("baselineValue"),
    collect_set("errorCode").alias("allErrorCodes"))
```

User-defined aggregation functions
> All user-defined aggregation functions are supported. See the Spark SQL programming guide (*https://oreil.ly/8nvJ2*) for more details on untyped and typed user-defined aggregation functions.

Regarding the execution of such streaming aggregations, we have already illustrated in previous sections how the running aggregates are maintained as a distributed state. In addition to this, there are two very important points to remember for aggregations not based on time: the output mode to use for such queries and planning the resource usage by state. These are discussed toward the end of this section. Next, we are going to discuss aggregations that combine data within time windows.

Aggregations with Event-Time Windows

In many cases, rather than running aggregations over the whole stream, you want aggregations over data bucketed by time windows. Continuing with our sensor example, say each sensor is expected to send at most one reading per minute and we want to detect if any sensor is reporting an unusually high number of times. To find such anomalies, we can count the number of readings received from each sensor in five-minute intervals. In addition, for robustness, we should be computing the time interval based on when the data was generated at the sensor and not based on when the data was received, as any transit delay would skew the results. In other words, we want to use the *event time*—that is, the timestamp in the record representing when the reading was generated. Say the sensorReadings DataFrame has the generation timestamp as a column named eventTime. We can express this five-minute count as follows:

```python
# In Python
from pyspark.sql.functions import *
(sensorReadings
  .groupBy("sensorId", window("eventTime", "5 minute"))
  .count())
```

```scala
// In Scala
import org.apache.spark.sql.functions.*
sensorReadings
  .groupBy("sensorId", window("eventTime", "5 minute"))
  .count()
```

The key thing to note here is the window() function, which allows us to express the five-minute windows as a dynamically computed grouping column. When started, this query will effectively do the following for each sensor reading:

- Use the eventTime value to compute the five-minute time window the sensor reading falls into.
- Group the reading based on the composite group (*<computed window>*, SensorId).
- Update the count of the composite group.

Let's understand this with an illustrative example. Figure 8-7 shows how a few sensor readings are mapped to groups of five-minute tumbling (i.e., nonoverlapping) windows based on their event time. The two timelines show when each received event will be processed by Structured Streaming, and the timestamp in the event data (usually, the time when the event was generated at the sensor).

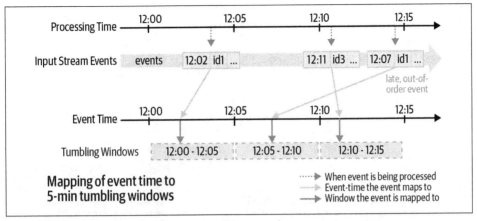

Figure 8-7. Mapping of event time to tumbling windows

Each five-minute window over event time is considered for the grouping based on which the counts will be calculated. Note that events may come late and out of order in terms of event time. As shown in the figure, the event with event time 12:07 was received and processed after the event with time 12:11. However, irrespective of when they arrive, each event is assigned to the appropriate group based on its event time. In fact, depending on the window specification, each event can be assigned to multiple groups. For example, if you want to compute counts corresponding to 10-minute windows sliding every 5 minutes, then you can do the following:

```python
# In Python
(sensorReadings
  .groupBy("sensorId", window("eventTime", "10 minute", "5 minute"))
  .count())
```

```scala
// In Scala
sensorReadings
  .groupBy("sensorId", window("eventTime", "10 minute", "5 minute"))
  .count()
```

In this query, every event will be assigned to two overlapping windows as illustrated in Figure 8-8.

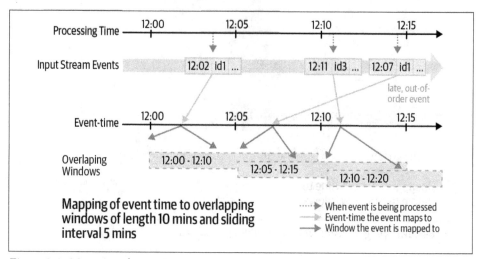

Figure 8-8. Mapping of event time to multiple overlapping windows

Each unique tuple of (`<assigned time window>`, `sensorId`) is considered a dynamically generated group for which counts will be computed. For example, the event [`eventTime = 12:07`, `sensorId = id1`] gets mapped to two time windows and therefore two groups, (`12:00-12:10`, `id1`) and (`12:05-12:15`, `id1`). The counts for these two windows are each incremented by 1. Figure 8-9 illustrates this for the previously shown events.

Assuming that the input records were processed with a trigger interval of five minutes, the tables at the bottom of Figure 8-9 show the state of the result table (i.e., the counts) at each of the micro-batches. As the event time moves forward, new groups are automatically created and their aggregates are automatically updated. Late and out-of-order events get handled automatically, as they simply update older groups.

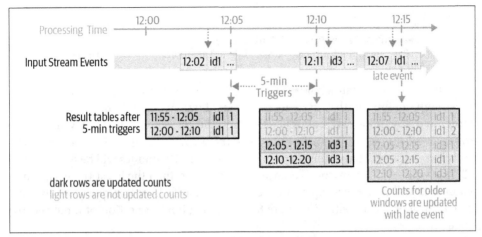

Figure 8-9. Updated counts in the result table after each five-minute trigger

However, from the point of view of resource usage, this poses a different problem—indefinitely growing state size. As new groups are created corresponding to the latest time windows, the older groups continue to occupy the state memory, waiting for any late data to update them. Even if in practice there is a bound on how late the input data can be (e.g., data cannot be more than seven days late), the query does not know that information. Hence, it does not know when to consider a window as "too old to receive updates" and drop it from the state. To provide a lateness bound to a query (and prevent unbounded state), you can specify *watermarks*, as we discuss next.

Handling late data with watermarks

A *watermark* is defined as a moving threshold in event time that trails behind the maximum event time seen by the query in the processed data. The trailing gap, known as the *watermark delay*, defines how long the engine will wait for late data to arrive. By knowing the point at which no more data will arrive for a given group, the engine can automatically finalize the aggregates of certain groups and drop them from the state. This limits the total amount of state that the engine has to maintain to compute the results of the query.

For example, suppose you know that your sensor data will not be late by more than 10 minutes. Then you can set the watermark as follows:

```
# In Python
(sensorReadings
  .withWatermark("eventTime", "10 minutes")
  .groupBy("sensorId", window("eventTime", "10 minutes", "5 minutes"))
  .mean("value"))
```

```scala
// In Scala
sensorReadings
  .withWatermark("eventTime", "10 minutes")
  .groupBy("sensorId", window("eventTime", "10 minutes", "5 minute"))
  .mean("value")
```

Note that you must call withWatermark() before the groupBy() and on the same timestamp column as that used to define windows. When this query is executed, Structured Streaming will continuously track the maximum observed value of the eventTime column and accordingly update the watermark, filter the "too late" data, and clear old state. That is, any data late by more than 10 minutes will be ignored, and all time windows that are more than 10 minutes older than the latest (by event time) input data will be cleaned up from the state. To clarify how this query will be executed, consider the timeline in Figure 8-10 showing how a selection of input records were processed.

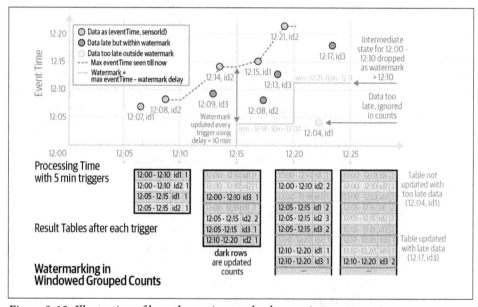

Figure 8-10. Illustration of how the engine tracks the maximum event time across events, updates the watermark, and accordingly handles late data

This figure shows a two-dimensional plot of records processed in terms of their processing times (x-axis) and their event times (y-axis). The records are processed in micro-batches of five minutes and marked with circles. The tables at the bottom show the state of the result table after each micro-batch completes.

Each record was received and processed after all the records to its left. Consider the two records [12:15, id1] (processed around 12:17) and [12:13, id3] (processed around 12:18). The record for id3 was considered late (and therefore marked in solid red) because it was generated by the sensor before the record for id1 but it was processed after the latter. However, in the micro-batch for processing-time range 12:15–12:20, the watermark used was 12:04 which was calculated based on the maximum event time seen till the previous micro-batch (that is, 12:14 minus the 10-minute watermark delay). Therefore, the late record [12:13, id3] was not considered to be too late and was successfully counted. In contrast, in the next micro-batch, the record [12:04, id1] was considered to be too late compared to the new watermark of 12:11 and was discarded.

You can set the watermark delay based on the requirements of your application—larger values for this parameter allow data to arrive later, but at the cost of increased state size (i.e., memory usage), and vice versa.

Semantic guarantees with watermarks.   Before we conclude this section about watermarks, let's consider the precise semantic guarantee that watermarking provides. A watermark of 10 minutes guarantees that the engine will *never drop any data* that is delayed by less than 10 minutes compared to the latest event time seen in the input data. However, the guarantee is strict only in one direction. Data delayed by more than 10 minutes is not guaranteed to be dropped—that is, it may get aggregated. Whether an input record more than 10 minutes late will actually be aggregated or not depends on the exact timing of when the record was received and when the micro-batch processing it was triggered.

Supported output modes

Unlike streaming aggregations not involving time, aggregations with time windows can use all three output modes. However, there are other implications regarding state cleanup that you need to be aware of, depending on the mode:

Update mode
    In this mode, every micro-batch will output only the rows where the aggregate got updated. This mode can be used with all types of aggregations. Specifically for time window aggregations, watermarking will ensure that the state will get cleaned up regularly. This is the most useful and efficient mode to run queries with streaming aggregations. However, you cannot use this mode to write aggregates to append-only streaming sinks, such as any file-based formats like Parquet and ORC (unless you use Delta Lake, which we will discuss in the next chapter).

Complete mode
    In this mode, every micro-batch will output all the updated aggregates, irrespective of their age or whether they contain changes. While this mode can be used on all types of aggregations, for time window aggregations, using complete mode means state will not be cleaned up even if a watermark is specified. Outputting all aggregates requires all past state, and hence aggregation data must be preserved

even if a watermark has been defined. Use this mode on time window aggregations with caution, as this can lead to an indefinite increase in state size and memory usage.

Append mode

This mode can be used only with aggregations on event-time windows and with watermarking enabled. Recall that append mode does not allow previously output results to change. For any aggregation without watermarks, every aggregate may be updated with any future data, and hence these cannot be output in append mode. Only when watermarking is enabled on aggregations on event-time windows does the query know when an aggregate is not going to update any further. Hence, instead of outputting the updated rows, append mode outputs each key and its final aggregate value only when the watermark ensures that the aggregate is not going to be updated again. The advantage of this mode is that it allows you to write aggregates to append-only streaming sinks (e.g., files). The disadvantage is that the output will be delayed by the watermark duration—the query has to wait for the trailing watermark to exceed the time window of a key before its aggregate can be finalized.

Streaming Joins

Structured Streaming supports joining a streaming Dataset with another static or streaming Dataset. In this section we will explore what types of joins (inner, outer, etc.) are supported, and how to use watermarks to limit the state stored for stateful joins. We will start with the simple case of joining a data stream and a static Dataset.

Stream–Static Joins

Many use cases require joining a data stream with a static Dataset. For example, let's consider the case of ad monetization. Suppose you are an advertisement company that shows ads on websites and you make money when users click on them. Let's assume that you have a static Dataset of all the ads to be shown (known as impressions), and another stream of events for each time users click on the displayed ads. To calculate the click revenue, you have to match each click in the event stream to the corresponding ad impression in the table. Let's first represent the data as two Data-Frames, a static one and a streaming one, as shown here:

```python
# In Python
# Static DataFrame [adId: String, impressionTime: Timestamp, ...]
# reading from your static data source
impressionsStatic = spark.read. ...

# Streaming DataFrame [adId: String, clickTime: Timestamp, ...]
# reading from your streaming source
clicksStream = spark.readStream. ...
```

```
// In Scala
// Static DataFrame [adId: String, impressionTime: Timestamp, ...]
// reading from your static data source
val impressionsStatic = spark.read. ...

// Streaming DataFrame [adId: String, clickTime: Timestamp, ...]
// reading from your streaming source
val clicksStream = spark.readStream. ...
```

To match the clicks with the impressions, you can simply apply an inner equi-join between them using the common adId column:

```
# In Python
matched = clicksStream.join(impressionsStatic, "adId")
```

```
// In Scala
val matched = clicksStream.join(impressionsStatic, "adId")
```

This is the same code as you would have written if both impressions and clicks were static DataFrames—the only difference is that you use spark.read() for batch processing and spark.readStream() for a stream. When this code is executed, every micro-batch of clicks is inner-joined against the static impression table to generate the output stream of matched events.

Besides inner joins, Structured Streaming also supports two types of stream–static outer joins:

- Left outer join when the left side is a streaming DataFrame
- Right outer join when the right side is a streaming DataFrame

The other kinds of outer joins (e.g., full outer and left outer with a streaming DataFrame on the right) are not supported because they are not easy to run incrementally. In both supported cases, the code is exactly as it would be for a left/right outer join between two static DataFrames:

```
# In Python
matched = clicksStream.join(impressionsStatic, "adId", "leftOuter")
```

```
// In Scala
val matched = clicksStream.join(impressionsStatic, Seq("adId"), "leftOuter")
```

There are a few key points to note about stream–static joins:

- Stream–static joins are stateless operations, and therefore do not require any kind of watermarking.
- The static DataFrame is read repeatedly while joining with the streaming data of every micro-batch, so you can cache the static DataFrame to speed up the reads.
- If the underlying data in the data source on which the static DataFrame was defined changes, whether those changes are seen by the streaming query depends

on the specific behavior of the data source. For example, if the static DataFrame was defined on files, then changes to those files (e.g., appends) will not be picked up until the streaming query is restarted.

In this stream–static example, we made a significant assumption: that the impression table is a static table. In reality, there will be a stream of new impressions generated as new ads are displayed. While stream–static joins are good for enriching data in one stream with additional static (or slowly changing) information, this approach is insufficient when both sources of data are changing rapidly. For that you need stream–stream joins, which we will discuss next.

Stream–Stream Joins

The challenge of generating joins between two data streams is that, at any point in time, the view of either Dataset is incomplete, making it much harder to find matches between inputs. The matching events from the two streams may arrive in any order and may be arbitrarily delayed. For example, in our advertising use case an impression event and its corresponding click event may arrive out of order, with arbitrary delays between them. Structured Streaming accounts for such delays by buffering the input data from both sides as the streaming state, and continuously checking for matches as new data is received. The conceptual idea is sketched out in Figure 8-11.

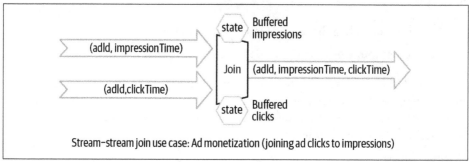

Figure 8-11. Ad monetization using a stream–stream join

Let's consider this in more detail, first with inner joins and then with outer joins.

Inner joins with optional watermarking

Say we have redefined our `impressions` DataFrame to be a streaming DataFrame. To get the stream of matching impressions and their corresponding clicks, we can use the same code we used earlier for static joins and stream–static joins:

```
# In Python
# Streaming DataFrame [adId: String, impressionTime: Timestamp, ...]
impressions = spark.readStream. ...

# Streaming DataFrame[adId: String, clickTime: Timestamp, ...]
clicks = spark.readStream. ...
matched = impressions.join(clicks, "adId")

// In Scala
// Streaming DataFrame [adId: String, impressionTime: Timestamp, ...]
val impressions = spark.readStream. ...

// Streaming DataFrame[adId: String, clickTime: Timestamp, ...]
val clicks = spark.readStream. ...
val matched = impressions.join(clicks, "adId")
```

Even though the code is the same, the execution is completely different. When this query is executed, the processing engine will recognize it to be a stream–stream join instead of a stream–static join. The engine will buffer all clicks and impressions as state, and will generate a matching impression-and-click as soon as a received click matches a buffered impression (or vice versa, depending on which was received first). Let's visualize how this inner join works using the example timeline of events in Figure 8-12.

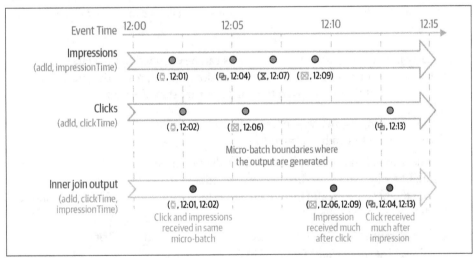

Figure 8-12. Illustrative timeline of clicks, impressions, and their joined output

In Figure 8-12, the blue dots represent the event times of impression and click events that were received across different micro-batches (separated by the dashed grey lines). For the purposes of this illustration, assume that each event was actually received at the same wall clock time as the event time. Note the different scenarios under which the related events are being joined. Both events with adId = 🖸 were received in the same micro-batch, so their joined output was generated by that micro-batch. However, for `adId` = ⊞ the impression was received at 12:04, much earlier than its corresponding click at 12:13. Structured Streaming will first receive the impression at 12:04 and buffer it in the state. For each received click, the engine will try to join it with all buffered impressions (and vice versa). Eventually, in a later micro-batch running around 12:13, the engine receives the click for `adId` = ⊞ and generates the joined output.

However, in this query, we have not given any indication of how long the engine should buffer an event to find a match. Therefore, the engine may buffer an event forever and accumulate an unbounded amount of streaming state. To limit the streaming state maintained by stream–stream joins, you need to know the following information about your use case:

- *What is the maximum time range between the generation of the two events at their respective sources?* In the context of our use case, let's assume that a click can occur within zero seconds to one hour after the corresponding impression.

- *What is the maximum duration an event can be delayed in transit between the source and the processing engine?* For example, ad clicks from a browser may get delayed due to intermittent connectivity and arrive much later than expected, and out of order. Let's say that impressions and clicks can be delayed by at most two and three hours, respectively.

These delay limits and event-time constraints can be encoded in the DataFrame operations using watermarks and time range conditions. In other words, you will have to do the following additional steps in the join to ensure state cleanup:

1. Define watermark delays on both inputs, such that the engine knows how delayed the input can be (similar to with streaming aggregations).

2. Define a constraint on event time across the two inputs, such that the engine can figure out when old rows of one input are not going to be required (i.e., will not satisfy the time constraint) for matches with the other input. This constraint can be defined in one of the following ways:

   a. Time range join conditions (e.g., join condition = `"leftTime BETWEEN rightTime AND rightTime + INTERVAL 1 HOUR"`)

   b. Join on event-time windows (e.g., join condition = `"leftTimeWindow = rightTimeWindow"`)

In our advertisement use case, our inner join code will get a little bit more complicated:

```
# In Python
# Define watermarks
impressionsWithWatermark = (impressions
  .selectExpr("adId AS impressionAdId", "impressionTime")
  .withWatermark("impressionTime", "2 hours"))

clicksWithWatermark = (clicks
  .selectExpr("adId AS clickAdId", "clickTime")
  .withWatermark("clickTime", "3 hours"))

# Inner join with time range conditions
(impressionsWithWatermark.join(clicksWithWatermark,
  expr("""
    clickAdId = impressionAdId AND
    clickTime BETWEEN impressionTime AND impressionTime + interval 1 hour""")))

// In Scala
// Define watermarks
val impressionsWithWatermark = impressions
  .selectExpr("adId AS impressionAdId", "impressionTime")
  .withWatermark("impressionTime", "2 hours ")

val clicksWithWatermark = clicks
  .selectExpr("adId AS clickAdId", "clickTime")
  .withWatermark("clickTime", "3 hours")

// Inner join with time range conditions
impressionsWithWatermark.join(clicksWithWatermark,
  expr("""
    clickAdId = impressionAdId AND
    clickTime BETWEEN impressionTime AND impressionTime + interval 1 hour"""))
```

With these time constraints for each event, the processing engine can automatically calculate how long events need to be buffered to generate correct results, and when the events can be dropped from the state. For example, it will evaluate the following (illustrated in Figure 8-13):

- Impressions need to be buffered for at most four hours (in event time), as a three-hour-late click may match with an impression made four hours ago (i.e., three hours late + up to one-hour delay between the impression and click).

- Conversely, clicks need to be buffered for at most two hours (in event time), as a two-hour-late impression may match with a click received two hours ago.

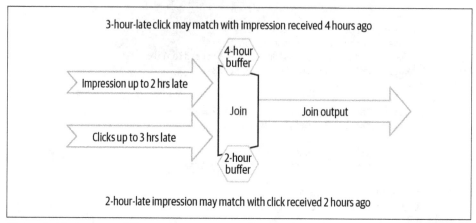

Figure 8-13. Structured Streaming automatically calculates thresholds for state cleanup using watermark delays and time range conditions

There are a few key points to remember about inner joins:

- For inner joins, specifying watermarking and event-time constraints are both optional. In other words, at the risk of potentially unbounded state, you may choose not to specify them. Only when both are specified will you get state cleanup.

- Similar to the guarantees provided by watermarking on aggregations, a watermark delay of two hours guarantees that the engine will never drop or not match any data that is less than two hours delayed, but data delayed by more than two hours may or may not get processed.

Outer joins with watermarking

The previous inner join will output only those ads for which both events have been received. In other words, ads that received no clicks will not be reported at all. Instead, you may want all ad impressions to be reported, with or without the associated click data, to enable additional analysis later (e.g., click-through rates). This brings us to *stream–stream outer joins*. All you need to do to implement this is specify the outer join type:

```
# In Python
# Left outer join with time range conditions
(impressionsWithWatermark.join(clicksWithWatermark,
  expr("""
    clickAdId = impressionAdId AND
    clickTime BETWEEN impressionTime AND impressionTime + interval 1 hour"""),
  "leftOuter"))  # only change: set the outer join type

// In Scala
// Left outer join with time range conditions
```

```
impressionsWithWatermark.join(clicksWithWatermark,
  expr("""
    clickAdId = impressionAdId AND
    clickTime BETWEEN impressionTime AND impressionTime + interval 1 hour"""),
  "leftOuter")  // Only change: set the outer join type
```

As expected of outer joins, this query will start generating output for every impression, with or without (i.e., using NULLs) the click data. However, there are a few additional points to note about outer joins:

- Unlike with inner joins, the watermark delay and event-time constraints are not optional for outer joins. This is because for generating the NULL results, the engine must know when an event is not going to match with anything else in the future. For correct outer join results and state cleanup, the watermarking and event-time constraints must be specified.

- Consequently, the outer NULL results will be generated with a delay as the engine has to wait for a while to ensure that there neither were nor would be any matches. This delay is the maximum buffering time (with respect to event time) calculated by the engine for each event as discussed in the previous section (i.e., four hours for impressions and two hours for clicks).

Arbitrary Stateful Computations

Many use cases require more complicated logic than the SQL operations we have discussed up to now. For example, say you want to track the statuses (e.g., signed in, busy, idle) of users by tracking their activities (e.g., clicks) in real time. To build this stream processing pipeline, you will have to track each user's activity history as a state with arbitrary data structure, and continuously apply arbitrarily complex changes on the data structure based on the user's actions. The operation mapGroupsWithState() and its more flexible counterpart flatMapGroupsWithState() are designed for such complex analytical use cases.

As of Spark 3.0, these two operations are only available in Scala and Java.

In this section, we will start with a simple example with mapGroupsWithState() to illustrate the four key steps to modeling custom state data and defining custom operations on it. Then we will discuss the concept of timeouts and how you can use them to expire state that has not been updated for a while. We will end with flatMapGroupsWithState(), which gives you even more flexibility.

Modeling Arbitrary Stateful Operations with mapGroupsWithState()

State with an arbitrary schema and arbitrary transformations on the state is modeled as a user-defined function that takes the previous version of the state value and new data as inputs, and generates the updated state and computed result as outputs. Programmatically in Scala, you will have to define a function with the following signature (K, V, S, and U are data types, as explained shortly):

```scala
// In Scala
def arbitraryStateUpdateFunction(
    key: K,
    newDataForKey: Iterator[V],
    previousStateForKey: GroupState[S]
): U
```

This function is provided to a streaming query using the operations `groupByKey()` and `mapGroupsWithState()`, as follows:

```scala
// In Scala
val inputDataset: Dataset[V] =  // input streaming Dataset

inputDataset
  .groupByKey(keyFunction)    // keyFunction() generates key from input
  .mapGroupsWithState(arbitraryStateUpdateFunction)
```

When this streaming query is started, in each micro-batch Spark will call this `arbitraryStateUpdateFunction()` for each unique key in the micro-batch's data. Let's take a closer look at what the parameters are and what parameter values Spark will call the function with:

key: K

> K is the data type of the common keys defined in the state and the input. Spark will call this function for each unique key in the data.

newDataForKey: Iterator[V]

> V is the data type of the input Dataset. When Spark calls this function for a key, this parameter will have all the new input data corresponding to that key. Note that the order in which the input data objects will be present in the iterator is not defined.

previousStateForKey: GroupState[S]

> S is the data type of the arbitrary state you are going to maintain, and Group State[S] is a typed wrapper object that provides methods to access and manage the state value. When Spark calls this function for a key, this object will provide the state value set the previous time Spark called this function for that key (i.e., for one of the previous micro-batches).

U

U is the data type of the output of the function.

> There are a couple of additional parameters that you have to pro-
> vide. All the types (K, V, S, U) must be encodable by Spark SQL's
> encoders. Accordingly, in mapGroupsWithState(), you have to pro-
> vide the typed encoders for S and U either implicitly in Scala or
> explicitly in Java. See "Dataset Encoders" on page 168 in Chapter 6
> for more details.

Let's examine how to express the desired state update function in this format with an
example. Say we want to understand user behavior based on their actions. Conceptually, it's quite simple: in every micro-batch, for each active user, we will use the new
actions taken by the user and update the user's "status." Programmatically, we can
define the state update function with the following steps:

1. Define the data types. We need to define the exact types of K, V, S, and U. In this
   case, we'll use the following:

   a. Input data (V) = case class UserAction(userId: String, action:
      String)

   b. Keys (K) = String (that is, the userId)

   c. State (S) = case class UserStatus(userId: String, active: Boolean)

   d. Output (U) = UserStatus, as we want to output the latest user status

   Note that all these data types are supported in encoders.

2. Define the function. Based on the chosen types, let's translate the conceptual idea
   into code. When this function is called with new user actions, there are two main
   situations we need to handle: whether a previous state (i.e., previous user status)
   exists for that key (i.e., userId) or not. Accordingly, we will initialize the user's
   status, or update the existing status with the new actions. We will explicitly
   update the state with the new running count, and finally return the updated
   userId-userStatus pair:

   ```
   // In Scala
   import org.apache.spark.sql.streaming._

   def updateUserStatus(
       userId: String,
       newActions: Iterator[UserAction],
       state: GroupState[UserStatus]): UserStatus = {

     val userStatus = state.getOption.getOrElse {
       new UserStatus(userId, false)
   ```

```
    }
    newActions.foreach { action =>
      userStatus.updateWith(action)
    }
    state.update(userStatus)
    return userStatus
  }
```

3. Apply the function on the actions. We will group the input actions Dataset using groupByKey() and then apply the updateUserStatus function using mapGroupsWithState():

```scala
// In Scala
val userActions: Dataset[UserAction] = ...
val latestStatuses = userActions
  .groupByKey(userAction => userAction.userId)
  .mapGroupsWithState(updateUserStatus _)
```

Once we start this streaming query with console output, we will see the updated user statuses being printed.

Before we move on to more advanced topics, there are a few notable points to remember:

- When the function is called, there is no well-defined order for the input records in the new data iterator (e.g., newActions). If you need to update the state with the input records in a specific order (e.g., in the order the actions were performed), then you have to explicitly reorder them (e.g., based on the event time-stamp or some other ordering ID). In fact, if there is a possibility that actions may be read out of order from the source, then you have to consider the possibility that a future micro-batch may receive data that should be processed before the data in the current batch. In that case, you have to buffer the records as part of the state.

- In a micro-batch, the function is called on a key once only if the micro-batch has data for that key. For example, if a user becomes inactive and provides no new actions for a long time, then by default, the function will not be called for a long time. If you want to update or remove state based on a user's inactivity over an extended period you have to use timeouts, which we will discuss in the next section.

- The output of mapGroupsWithState() is assumed by the incremental processing engine to be continuously updated key/value records, similar to the output of aggregations. This limits what operations are supported in the query after mapGroupsWithState(), and what sinks are supported. For example, appending the output into files is not supported. If you want to apply arbitrary stateful

operations with greater flexibility, then you have to use `flatMapGroupsWith`
`State()`. We will discuss that after timeouts.

Using Timeouts to Manage Inactive Groups

In the preceding example of tracking active user sessions, as more users become active, the number of keys in the state will keep increasing, and so will the memory used by the state. Now, in a real-world scenario, users are likely not going to stay active all the time. It may not be very useful to keep the status of inactive users in the state, as it is not going to change again until those users become active again. Hence, we may want to explicitly drop all information for inactive users. However, a user may not explicitly take any action to become inactive (e.g., explicitly logging off), and we may have to define inactivity as lack of any action for a threshold duration. This becomes tricky to encode in the function, as the function is not called for a user until there are new actions from that user.

To encode time-based inactivity, `mapGroupsWithState()` supports timeouts that are defined as follows:

- Each time the function is called on a key, a timeout can be set on the key based on a duration or a threshold timestamp.
- If that key does not receive any data, such that the timeout condition is met, the key is marked as "timed out." The next micro-batch will call the function on this timed-out key even if there is no data for that key in that micro-batch. In this special function call, the new input data iterator will be empty (since there is no new data) and `GroupState.hasTimedOut()` will return `true`. This is the best way to identify inside the function whether the call was due to new data or a timeout.

There are two types of timeouts, based on our two notions of time: processing time and event time. The processing-time timeout is the simpler of the two to use, so we'll start with that.

Processing-time timeouts

Processing-time timeouts are based on the system time (also known as the wall clock time) of the machine running the streaming query and are defined as follows: if a key last received data at system timestamp T, and the current timestamp is more than (T + `<timeout duration>`), then the function will be called again with a new empty data iterator.

Let's investigate how to use timeouts by updating our user example to remove a user's state based on one hour of inactivity. We will make three changes:

- In `mapGroupsWithState()`, we will specify the timeout as `GroupStateTime out.ProcessingTimeTimeout`.

- In the state update function, before updating the state with new data, we have to check whether the state has timed out or not. Accordingly, we will update or remove the state.

- In addition, every time we update the state with new data, we will set the timeout duration.

Here's the updated code:

```scala
// In Scala
def updateUserStatus(
    userId: String,
    newActions: Iterator[UserAction],
    state: GroupState[UserStatus]): UserStatus = {

  if (!state.hasTimedOut) {            // Was not called due to timeout
    val userStatus = state.getOption.getOrElse {
      new UserStatus(userId, false)
    }
    newActions.foreach { action => userStatus.updateWith(action) }
    state.update(userStatus)
    state.setTimeoutDuration("1 hour") // Set timeout duration
    return userStatus

  } else {
    val userStatus = state.get()
    state.remove()                     // Remove state when timed out
    return userStatus.asInactive()     // Return inactive user's status
  }
}

val latestStatuses = userActions
  .groupByKey(userAction => userAction.userId)
  .mapGroupsWithState(
    GroupStateTimeout.ProcessingTimeTimeout)(
    updateUserStatus _)
```

This query will automatically clean up the state of users for whom the query has not processed any data for more than an hour. However, there are a few points to note about timeouts:

- The timeout set by the last call to the function is automatically cancelled when the function is called again, either for the new received data or for the timeout. Hence, whenever the function is called, the timeout duration or timestamp needs to be explicitly set to enable the timeout.

- Since the timeouts are processed during the micro-batches, the timing of their execution is imprecise and depends heavily on the trigger interval and micro-batch processing times. Therefore, it is not advised to use timeouts for precise timing control.

- While processing-time timeouts are simple to reason about, they are not robust to slowdowns and downtimes. If the streaming query suffers a downtime of more than one hour, then after restart, all the keys in the state will be timed out because more than one hour has passed since each key received data. Similar wide-scale timeouts can occur if the query processes data slower than it is arriving at the source (e.g., if data is arriving and getting buffered in Kafka). For example, if the timeout is five minutes, then a sudden drop in processing rate (or spike in data arrival rate) that causes a five-minute lag could produce spurious timeouts. To avoid such issues we can use an event-time timeout, which we will discuss next.

Event-time timeouts

Instead of the system clock time, an event-time timeout is based on the event time in the data (similar to time-based aggregations) and a watermark defined on the event time. If a key is configured with a specific timeout timestamp of T (i.e., not a duration), then that key will time out when the watermark exceeds T if no new data was received for that key since the last time the function was called. Recall that the watermark is a moving threshold that lags behind the maximum event time seen while processing the data. Hence, unlike system time, the watermark moves forward in time at the same rate as the data is processed. This means (unlike with processing-time timeouts) any slowdown or downtime in query processing will not cause spurious timeouts.

Let's modify our example to use an event-time timeout. In addition to the changes we already made for using the processing-time timeout, we will make the following changes:

- Define watermarks on the input Dataset (assume that the class UserAction has an eventTimestamp field). Recall that the watermark threshold represents the acceptable amount of time by which input data can be late and out of order.

- Update mapGroupsWithState() to use EventTimeTimeout.

- Update the function to set the threshold timestamp at which the timeout will occur. Note that event-time timeouts do not allow setting a timeout duration, like processing-time timeouts. We will discuss the reason for this later. In this example, we will calculate this timeout as the current watermark plus one hour.

Here is the updated example:

```scala
// In Scala
def updateUserStatus(
    userId: String,
    newActions: Iterator[UserAction],
    state: GroupState[UserStatus]):UserStatus = {

  if (!state.hasTimedOut) {  // Was not called due to timeout
    val userStatus = if (state.getOption.getOrElse {
      new UserStatus()
    }
    newActions.foreach { action => userStatus.updateWith(action) }
    state.update(userStatus)

    // Set the timeout timestamp to the current watermark + 1 hour
    state.setTimeoutTimestamp(state.getCurrentWatermarkMs, "1 hour")
    return userStatus
  } else {
    val userStatus = state.get()
    state.remove()
    return userStatus.asInactive() }
}

val latestStatuses = userActions
  .withWatermark("eventTimestamp", "10 minutes")
  .groupByKey(userAction => userAction.userId)
  .mapGroupsWithState(
    GroupStateTimeout.EventTimeTimeout)(
    updateUserStatus _)
```

This query will be much more robust to spurious timeouts caused by restarts and processing delays.

Here are a few points to note about event-time timeouts:

- Unlike in the previous example with processing-time timeouts, we have used `GroupState.setTimeoutTimestamp()` instead of `GroupState.setTimeoutDura tion()`. This is because with processing-time timeouts the duration is sufficient to calculate the exact future timestamp (i.e., current system time + specified duration) when the timeout would occur, but this is not the case for event-time timeouts. Different applications may want to use different strategies to calculate the threshold timestamp. In this example we simply calculate it based on the current watermark, but a different application may instead choose to calculate a key's timeout timestamp based on the maximum event-time timestamp seen for that key (tracked and saved as part of the state).

- The timeout timestamp must be set to a value larger than the current watermark. This is because the timeout is expected to happen when the timestamp crosses the watermark, so it's illogical to set the timestamp to a value already larger than the current watermark.

Before we move on from timeouts, one last thing to remember is that you can use these timeout mechanisms for more creative processing than fixed-duration timeouts. For example, you can implement an approximately periodic task (say, every hour) on the state by saving the last task execution timestamp in the state and using that to set the processing-time timeout duration, as shown in this code snippet:

```scala
// In Scala
timeoutDurationMs = lastTaskTimstampMs + periodIntervalMs -
groupState.getCurrentProcessingTimeMs()
```

Generalization with flatMapGroupsWithState()

There are two key limitations with `mapGroupsWithState()` that may limit the flexibility that we want to implement more complex use cases (e.g., chained sessionizations):

- Every time `mapGroupsWithState()` is called, you have to return one and only one record. For some applications, in some triggers, you may not want to output anything at all.

- With `mapGroupsWithState()`, due to the lack of more information about the opaque state update function, the engine assumes that generated records are updated key/value data pairs. Accordingly, it reasons about downstream operations and allows or disallows some of them. For example, the DataFrame generated using `mapGroupsWithState()` cannot be written out in append mode to files. However, some applications may want to generate records that can be considered as appends.

`flatMapGroupsWithState()` overcomes these limitations, at the cost of slightly more complex syntax. It has two differences from `mapGroupsWithState()`:

- The return type is an iterator, instead of a single object. This allows the function to return any number of records, or, if needed, no records at all.

- It takes another parameter, called the *operator output mode* (not to be confused with the query output modes we discussed earlier in the chapter), that defines whether the output records are new records that can be appended (`Output Mode.Append`) or updated key/value records (`OutputMode.Update`).

To illustrate the use of this function, let's extend our user tracking example (we have removed timeouts to keep the code simple). For example, if we want to generate alerts only for certain user changes and we want to write the output alerts to files, we can do the following:

```scala
// In Scala
def getUserAlerts(
    userId: String,
    newActions: Iterator[UserAction],
```

```
    state: GroupState[UserStatus]): Iterator[UserAlert] = {

  val userStatus = state.getOption.getOrElse {
    new UserStatus(userId, false)
  }
  newActions.foreach { action =>
    userStatus.updateWith(action)
  }
  state.update(userStatus)

  // Generate any number of alerts
  return userStatus.generateAlerts().toIterator
}

val userAlerts = userActions
  .groupByKey(userAction => userAction.userId)
  .flatMapGroupsWithState(
    OutputMode.Append,
    GroupStateTimeout.NoTimeout)(
    getUserAlerts)
```

Performance Tuning

Structured Streaming uses the Spark SQL engine and therefore can be tuned with the same parameters as those discussed for Spark SQL in Chapters 5 and 7. However, unlike batch jobs that may process gigabytes to terabytes of data, micro-batch jobs usually process much smaller volumes of data. Hence, a Spark cluster running streaming queries usually needs to be tuned slightly differently. Here are a few considerations to keep in mind:

Cluster resource provisioning

Since Spark clusters running streaming queries are going to run 24/7, it is important to provision resources appropriately. Underprovisoning the resources can cause the streaming queries to fall behind (with micro-batches taking longer and longer), while overprovisioning (e.g., allocated but unused cores) can cause unnecessary costs. Furthermore, allocation should be done based on the nature of the streaming queries: stateless queries usually need more cores, and stateful queries usually need more memory.

Number of partitions for shuffles

For Structured Streaming queries, the number of shuffle partitions usually needs to be set much lower than for most batch queries—dividing the computation too much increases overheads and reduces throughput. Furthermore, shuffles due to stateful operations have significantly higher task overheads due to checkpointing. Hence, for streaming queries with stateful operations and trigger intervals of a few seconds to minutes, it is recommended to tune the number of shuffle partitions from the default value of 200 to at most two to three times the number of allocated cores.

Setting source rate limits for stability

After the allocated resources and configurations have been optimized for a query's expected input data rates, it's possible that sudden surges in data rates can generate unexpectedly large jobs and subsequent instability. Besides the costly approach of overprovisioning, you can safeguard against instability using source rate limits. Setting limits in supported sources (e.g., Kafka and files) prevents a query from consuming too much data in a single micro-batch. The surge data will stay buffered in the source, and the query will eventually catch up. However, note the following:

- Setting the limit too low can cause the query to underutilize allocated resources and fall behind the input rate.

- Limits do not effectively guard against sustained increases in input rate. While stability is maintained, the volume of buffered, unprocessed data will grow indefinitely at the source and so will the end-to-end latencies.

Multiple streaming queries in the same Spark application

Running multiple streaming queries in the same `SparkContext` or `SparkSession` can lead to fine-grained resource sharing. However:

- Executing each query continuously uses resources in the Spark driver (i.e., the JVM where it is running). This limits the number of queries that the driver can execute simultaneously. Hitting those limits can either bottleneck the task scheduling (i.e., underutilizing the executors) or exceed memory limits.

- You can ensure fairer resource allocation between queries in the same context by setting them to run in separate scheduler pools. Set the `SparkContext`'s thread-local property `spark.scheduler.pool` to a different string value for each stream:

```
// In Scala
// Run streaming query1 in scheduler pool1
spark.sparkContext.setLocalProperty("spark.scheduler.pool", "pool1")
df.writeStream.queryName("query1").format("parquet").start(path1)

// Run streaming query2 in scheduler pool2
spark.sparkContext.setLocalProperty("spark.scheduler.pool", "pool2")
df.writeStream.queryName("query2").format("parquet").start(path2)
```

```
# In Python
# Run streaming query1 in scheduler pool1
spark.sparkContext.setLocalProperty("spark.scheduler.pool", "pool1")
df.writeStream.queryName("query1").format("parquet").start(path1)

# Run streaming query2 in scheduler pool2
spark.sparkContext.setLocalProperty("spark.scheduler.pool", "pool2")
df.writeStream.queryName("query2").format("parquet").start(path2)
```

Summary

This chapter explored writing Structured Streaming queries using the DataFrame API. Specifically, we discussed:

- The central philosophy of Structured Streaming and the processing model of treating input data streams as unbounded tables
- The key steps to define, start, restart, and monitor streaming queries
- How to use various built-in streaming sources and sinks and write custom streaming sinks
- How to use and tune managed stateful operations like streaming aggregations and stream–stream joins
- Techniques for expressing custom stateful computations

By working through the code snippets in the chapter and the notebooks in the book's GitHub repo (*https://github.com/databricks/LearningSparkV2*), you will get a feel for how to use Structured Streaming effectively. In the next chapter, we explore how you can manage structured data read and written simultaneously from batch and streaming workloads.

Building Reliable Data Lakes
with Apache Spark

In the previous chapters, you learned how to easily and effectively use Apache Spark to build scalable and performant data processing pipelines. However, in practice, expressing the processing logic only solves half of the end-to-end problem of building a pipeline. For a data engineer, data scientist, or data analyst, the ultimate goal of building pipelines is to query the processed data and get insights from it. The choice of storage solution determines the end-to-end (i.e., from raw data to insights) robustness and performance of the data pipeline.

In this chapter, we will first discuss the key features of a storage solution that you need to look out for. Then we will discuss two broad classes of storage solutions, databases and data lakes, and how to use Apache Spark with them. Finally, we will introduce the next wave of storage solution, called lakehouses, and explore some of the new open source processing engines in this space.

The Importance of an Optimal Storage Solution

Here are some of the properties that are desired in a storage solution:

Scalability and performance
> The storage solution should be able to scale to the volume of data and provide the read/write throughput and latency that the workload requires.

Transaction support
> Complex workloads are often reading and writing data concurrently, so support for ACID transactions (*https://oreil.ly/6Jn97*) is essential to ensure the quality of the end results.

Support for diverse data formats

The storage solution should be able to store unstructured data (e.g., text files like raw logs), semi-structured data (e.g., JSON data), and structured data (e.g., tabular data).

Support for diverse workloads

The storage solution should be able to support a diverse range of business workloads, including:

- SQL workloads like traditional BI analytics
- Batch workloads like traditional ETL jobs processing raw unstructured data
- Streaming workloads like real-time monitoring and alerting
- ML and AI workloads like recommendations and churn predictions

Openness

Supporting a wide range of workloads often requires the data to be stored in open data formats. Standard APIs allow the data to be accessed from a variety of tools and engines. This allows the business to use the most optimal tools for each type of workload and make the best business decisions.

Over time, different kinds of storage solutions have been proposed, each with its unique advantages and disadvantages with respect to these properties. In this chapter, we will explore how the available storage solutions evolved from *databases* to *data lakes*, and how to use Apache Spark with each of them. We'll then turn our attention to the next generation of storage solutions, often called data *lakehouses*, that can provide the best of both worlds: the scalability and flexibility of data lakes with the transactional guarantees of databases.

Databases

For many decades, databases have been the most reliable solution for building data warehouses to store business-critical data. In this section, we will explore the architecture of databases and their workloads, and how to use Apache Spark for analytics workloads on databases. We will end this section with a discussion of the limitations of databases in supporting modern non-SQL workloads.

A Brief Introduction to Databases

Databases are designed to store structured data as tables, which can be read using SQL queries. The data must adhere to a strict schema, which allows a database management system to heavily co-optimize the data storage and processing. That is, they tightly couple their internal layout of the data and indexes in on-disk files with their highly optimized query processing engines, thus providing very fast computations on the stored data along with strong transactional ACID guarantees on all read/write operations.

SQL workloads on databases can be broadly classified into two categories, as follows:

Online transaction processing (OLTP) workloads (https://oreil.ly/n94tD)
> Like bank account transactions, OLTP workloads are typically high-concurrency, low-latency, simple queries that read or update a few records at a time.

Online analytical processing (OLAP) (https://oreil.ly/NJQ2m)
> OLAP workloads, like periodic reporting, are typically complex queries (involving aggregates and joins) that require high-throughput scans over many records.

It is important to note that Apache Spark is a query engine that is primarily designed for OLAP workloads, not OLTP workloads. Hence, in the rest of the chapter we are going to focus our discussion on storage solutions for analytical workloads. Next, let's see how Apache Spark can be used to read from and write to databases.

Reading from and Writing to Databases Using Apache Spark

Thanks to the ever-growing ecosystem of connectors, Apache Spark can connect to a wide variety of databases for reading and writing data. For databases that have JDBC drivers (e.g., PostgreSQL, MySQL), you can use the built-in JDBC data source along with the appropriate JDBC driver jars to access the data. For many other modern databases (e.g., Azure Cosmos DB, Snowflake), there are dedicated connectors that you can invoke using the appropriate format name. Several examples were discussed in detail in Chapter 5. This makes it very easy to augment your data warehouses and databases with workloads and use cases based on Apache Spark.

Limitations of Databases

Since the last century, databases and SQL queries have been known as great building solutions for BI workloads. However, the last decade has seen two major new trends in analytical workloads:

Growth in data sizes
> With the advent of big data, there has been a global trend in the industry to measure and collect everything (page views, clicks, etc.) in order to understand trends and user behaviors. As a result, the amount of data collected by any company or organization has increased from gigabytes a couple of decades ago to terabytes and petabytes today.

Growth in the diversity of analytics
> Along with the increase in data collection, there is a need for deeper insights. This has led to an explosive growth of complex analytics like machine learning and deep learning.

Databases have been shown to be rather inadequate at accommodating these new trends, because of the following limitations:

Databases are extremely expensive to scale out
Although databases are extremely efficient at processing data on a single machine, the rate of growth of data volumes has far outpaced the growth in performance capabilities of a single machine. The only way forward for processing engines is to scale out—that is, use multiple machines to process data in parallel. However, most databases, especially the open source ones, are not designed for scaling out to perform distributed processing. The few industrial database solutions that can remotely keep up with the processing requirements tend to be proprietary solutions running on specialized hardware, and are therefore very expensive to acquire and maintain.

Databases do not support non–SQL based analytics very well
Databases store data in complex (often proprietary) formats that are typically highly optimized for only that database's SQL processing engine to read. This means other processing tools, like machine learning and deep learning systems, cannot efficiently access the data (except by inefficiently reading all the data from the database). Nor can databases be easily extended to perform non–SQL based analytics like machine learning.

These limitations of databases led to the development of a completely different approach to storing data, known as *data lakes*.

Data Lakes

In contrast to most databases, a data lake is a distributed storage solution that runs on commodity hardware and easily scales out horizontally. In this section, we will start with a discussion of how data lakes satisfy the requirements of modern workloads, then see how Apache Spark integrates with data lakes to make workloads scale to data of any size. Finally, we will explore the impact of the architectural sacrifices made by data lakes to achieve scalability.

A Brief Introduction to Data Lakes

The data lake architecture, unlike that of databases, decouples the distributed storage system from the distributed compute system. This allows each system to scale out as needed by the workload. Furthermore, the data is saved as files with open formats, such that any processing engine can read and write them using standard APIs. This idea was popularized in the late 2000s by the Hadoop File System (HDFS) from the Apache Hadoop project (*https://hadoop.apache.org/*), which itself was heavily inspired by the research paper "The Google File System" (*https://oreil.ly/v6py_*) by Sanjay Ghemawat, Howard Gobioff, and Shun-Tak Leung.

Organizations build their data lakes by independently choosing the following:

Storage system
> They choose to either run HDFS on a cluster of machines or use any cloud object store (e.g., AWS S3, Azure Data Lake Storage, or Google Cloud Storage).

File format
> Depending on the downstream workloads, the data is stored as files in either structured (e.g., Parquet, ORC), semi-structured (e.g., JSON), or sometimes even unstructured formats (e.g., text, images, audio, video).

Processing engine(s)
> Again, depending on the kinds of analytical workloads to be performed, a processing engine is chosen. This can either be a batch processing engine (e.g., Spark, Presto, Apache Hive), a stream processing engine (e.g., Spark, Apache Flink), or a machine learning library (e.g., Spark MLlib, scikit-learn, R).

This flexibility—the ability to choose the storage system, open data format, and processing engine that are best suited to the workload at hand—is the biggest advantage of data lakes over databases. On the whole, for the same performance characteristics, data lakes often provide a much cheaper solution than databases. This key advantage has led to the explosive growth of the big data ecosystem. In the next section, we will discuss how you can use Apache Spark to read and write common file formats on any storage system.

Reading from and Writing to Data Lakes using Apache Spark

Apache Spark is one of the best processing engines to use when building your own data lake, because it provides all the key features they require:

Support for diverse workloads
> Spark provides all the necessary tools to handle a diverse range of workloads, including batch processing, ETL operations, SQL workloads using Spark SQL, stream processing using Structured Streaming (discussed in Chapter 8), and machine learning using MLlib (discussed in Chapter 10), among many others.

Support for diverse file formats
> In Chapter 4, we explored in detail how Spark has built-in support for unstructured, semi-structured, and structured file formats.

Support for diverse filesystems
> Spark supports accessing data from any storage system that supports Hadoop's FileSystem APIs. Since this API has become the de facto standard in the big data ecosystem, most cloud and on-premises storage systems provide implementations for it—which means Spark can read from and write to most storage systems.

However, for many filesystems (especially those based on cloud storage, like AWS S3), you have to configure Spark such that it can access the filesystem in a secure manner. Furthermore, cloud storage systems often do not have the same file operation semantics expected from a standard filesystem (e.g., eventual consistency in S3), which can lead to inconsistent results if you do not configure Spark accordingly. See the documentation on cloud integration (*https://oreil.ly/YncTL*) for details.

Limitations of Data Lakes

Data lakes are not without their share of flaws, the most egregious of which is the lack of transactional guarantees. Specifically, data lakes fail to provide ACID guarantees on:

Atomicity and isolation

Processing engines write data in data lakes as many files in a distributed manner. If the operation fails, there is no mechanism to roll back the files already written, thus leaving behind potentially corrupted data (the problem is exacerbated when concurrent workloads modify the data because it is very difficult to provide isolation across files without higher-level mechanisms).

Consistency

Lack of atomicity on failed writes further causes readers to get an inconsistent view of the data. In fact, it is hard to ensure data quality even in successfully written data. For example, a very common issue with data lakes is accidentally writing out data files in a format and schema inconsistent with existing data.

To work around these limitations of data lakes, developers employ all sorts of tricks. Here are a few examples:

- Large collections of data files in data lakes are often "partitioned" by subdirectories based on a column's value (e.g., a large Parquet-formatted Hive table partitioned by date). To achieve atomic modifications of existing data, often entire subdirectories are rewritten (i.e., written to a temporary directory, then references swapped) just to update or delete a few records.

- The schedules of data update jobs (e.g., daily ETL jobs) and data querying jobs (e.g., daily reporting jobs) are often staggered to avoid concurrent access to the data and any inconsistencies caused by it.

Attempts to eliminate such practical issues have led to the development of new systems, such as lakehouses.

Lakehouses: The Next Step in the Evolution of Storage Solutions

The *lakehouse* is a new paradigm that combines the best elements of data lakes and data warehouses for OLAP workloads. Lakehouses are enabled by a new system design that provides data management features similar to databases directly on the low-cost, scalable storage used for data lakes. More specifically, they provide the following features:

Transaction support
> Similar to databases, lakehouses provide ACID guarantees in the presence of concurrent workloads.

Schema enforcement and governance
> Lakehouses prevent data with an incorrect schema being inserted into a table, and when needed, the table schema can be explicitly evolved to accommodate ever-changing data. The system should be able to reason about data integrity, and it should have robust governance and auditing mechanisms.

Support for diverse data types in open formats
> Unlike databases, but similar to data lakes, lakehouses can store, refine, analyze, and access all types of data needed for many new data applications, be it structured, semi-structured, or unstructured. To enable a wide variety of tools to access it directly and efficiently, the data must be stored in open formats with standardized APIs to read and write them.

Support for diverse workloads
> Powered by the variety of tools reading data using open APIs, lakehouses enable diverse workloads to operate on data in a single repository. Breaking down isolated data silos (i.e., multiple repositories for different categories of data) enables developers to more easily build diverse and complex data solutions, from traditional SQL and streaming analytics to machine learning.

Support for upserts and deletes
> Complex use cases like change-data-capture (CDC) (*https://oreil.ly/eEj_m*) and slowly changing dimension (SCD) (*https://oreil.ly/13zll*) operations require data in tables to be continuously updated. Lakehouses allow data to be concurrently deleted and updated with transactional guarantees.

Data governance
> Lakehouses provide the tools with which you can reason about data integrity and audit all the data changes for policy compliance.

Currently, there are a few open source systems, such as Apache Hudi, Apache Iceberg, and Delta Lake, that can be used to build lakehouses with these properties. At a very high level, all three projects have a similar architecture inspired by well-known database principles. They are all open data storage formats that do the following:

- Store large volumes of data in structured file formats on scalable filesystems.
- Maintain a transaction log to record a timeline of atomic changes to the data (much like databases).
- Use the log to define versions of the table data and provide snapshot isolation guarantees between readers and writers.
- Support reading and writing to tables using Apache Spark.

Within these broad strokes, each project has unique characteristics in terms of APIs, performance, and the level of integration with Apache Spark's data source APIs. We will explore them next. Note that all of these projects are evolving fast, and therefore some of the descriptions may be outdated at the time you are reading them. Refer to the online documentation for each project for the most up-to-date information.

Apache Hudi

Initially built by Uber Engineering (*https://eng.uber.com/hoodie*), Apache Hudi (*https://hudi.apache.org*)—an acronym for Hadoop Update Delete and Incremental— is a data storage format that is designed for incremental upserts and deletes over key/ value-style data. The data is stored as a combination of columnar formats (e.g., Parquet files) and row-based formats (e.g., Avro files for recording incremental changes over Parquet files). Besides the common features mentioned earlier, it supports:

- Upserting with fast, pluggable indexing
- Atomic publishing of data with rollback support
- Reading incremental changes to a table
- Savepoints for data recovery
- File size and layout management using statistics
- Async compaction of row and columnar data

Apache Iceberg

Originally built at Netflix (*https://github.com/Netflix/iceberg*), Apache Iceberg (*https:// iceberg.apache.org*) is another open storage format for huge data sets. However, unlike Hudi, which focuses on upserting key/value data, Iceberg focuses more on general-purpose data storage that scales to petabytes in a single table and has schema evolution properties. Specifically, it provides the following additional features (besides the common ones):

- Schema evolution by adding, dropping, updating, renaming, and reordering of columns, fields, and/or nested structures

- Hidden partitioning, which under the covers creates the partition values for rows in a table

- Partition evolution, where it automatically performs a metadata operation to update the table layout as data volume or query patterns change

- Time travel, which allows you to query a specific table snapshot by ID or by timestamp

- Rollback to previous versions to correct errors

- Serializable isolation, even between multiple concurrent writers

Delta Lake

Delta Lake (*https://delta.io/*) is an open source project hosted by the Linux Foundation, built by the original creators of Apache Spark. Similar to the others, it is an open data storage format that provides transactional guarantees and enables schema enforcement and evolution. It also provides several other interesting features, some of which are unique. Delta Lake supports:

- Streaming reading from and writing to tables using Structured Streaming sources and sinks

- Update, delete, and merge (for upserts) operations, even in Java, Scala, and Python APIs

- Schema evolution either by explicitly altering the table schema or by implicitly merging a DataFrame's schema to the table's during the DataFrame's write. (In fact, the merge operation in Delta Lake supports advanced syntax for conditional updates/inserts/deletes, updating all columns together, etc., as you'll see later in the chapter.)

- Time travel, which allows you to query a specific table snapshot by ID or by timestamp

- Rollback to previous versions to correct errors

- Serializable isolation between multiple concurrent writers performing any SQL, batch, or streaming operations

In the rest of this chapter, we are going to explore how such a system, along with Apache Spark, can be used to build a lakehouse that provides the aforementioned properties. Of these three systems, so far Delta Lake has the tightest integration with Apache Spark data sources (both for batch and streaming workloads) and SQL operations (e.g., MERGE). Hence, we will use Delta Lake as the vehicle for further exploration.

This project is called Delta Lake because of its analogy to streaming. Streams flow into the sea to create deltas—this is where all of the sediments accumulate, and thus where the valuable crops are grown. Jules S. Damji (one of our coauthors) came up with this!

Building Lakehouses with Apache Spark and Delta Lake

In this section, we are going to take a quick look at how Delta Lake and Apache Spark can be used to build lakehouses. Specifically, we will explore the following:

- Reading and writing Delta Lake tables using Apache Spark
- How Delta Lake allows concurrent batch and streaming writes with ACID guarantees
- How Delta Lake ensures better data quality by enforcing schema on all writes, while allowing for explicit schema evolution
- Building complex data pipelines using update, delete, and merge operations, all of which ensure ACID guarantees
- Auditing the history of operations that modified a Delta Lake table and traveling back in time by querying earlier versions of the table

The data we will use in this section is a modified version (a subset of columns in Parquet format) of the public Lending Club Loan Data (*https://oreil.ly/P7AR-*).[1] It includes all funded loans from 2012 to 2017. Each loan record includes applicant information provided by the applicant as well as the current loan status (current, late, fully paid, etc.) and latest payment information.

Configuring Apache Spark with Delta Lake

You can configure Apache Spark to link to the Delta Lake library in one of the following ways:

Set up an interactive shell

If you're using Apache Spark 3.0, you can start a PySpark or Scala shell with Delta Lake by using the following command-line argument:

```
--packages io.delta:delta-core_2.12:0.7.0
```

For example:

```
pyspark --packages io.delta:delta-core_2.12:0.7.0
```

If you are running Spark 2.4, you have to use Delta Lake 0.6.0.

---

1 A full view of the data is available at this Excel file (*https://oreil.ly/Rgtn1*).

Set up a standalone Scala/Java project using Maven coordinates

If you want to build a project using Delta Lake binaries from the Maven Central repository, you can add the following Maven coordinates to the project dependencies:

```
<dependency>
  <groupId>io.delta</groupId>
  <artifactId>delta-core_2.12</artifactId>
  <version>0.7.0</version>
</dependency>
```

Again, if you are running Spark 2.4 you have to use Delta Lake 0.6.0.

 See the Delta Lake documentation (*https://oreil.ly/MmlC3*) for the most up-to-date information.

Loading Data into a Delta Lake Table

If you are used to building data lakes with Apache Spark and any of the structured data formats—say, Parquet—then it is very easy to migrate existing workloads to use the Delta Lake format. All you have to do is change all the DataFrame read and write operations to use `format("delta")` instead of `format("parquet")`. Let's try this out with some of the aforementioned loan data, which is available as a Parquet file (*https://oreil.ly/7pP1y*). First let's read this data and save it as a Delta Lake table:

```
// In Scala
// Configure source data path
val sourcePath = "/databricks-datasets/learning-spark-v2/loans/
  loan-risks.snappy.parquet"

// Configure Delta Lake path
val deltaPath = "/tmp/loans_delta"

// Create the Delta table with the same loans data
spark
  .read
  .format("parquet")
  .load(sourcePath)
  .write
  .format("delta")
  .save(deltaPath)

// Create a view on the data called loans_delta
spark
  .read
  .format("delta")
```

```
  .load(deltaPath)
  .createOrReplaceTempView("loans_delta")

# In Python
# Configure source data path
sourcePath = "/databricks-datasets/learning-spark-v2/loans/
  loan-risks.snappy.parquet"

# Configure Delta Lake path
deltaPath = "/tmp/loans_delta"

# Create the Delta Lake table with the same loans data
(spark.read.format("parquet").load(sourcePath)
  .write.format("delta").save(deltaPath))

# Create a view on the data called loans_delta
spark.read.format("delta").load(deltaPath).createOrReplaceTempView("loans_delta")
```

Now we can read and explore the data as easily as any other table:

```
// In Scala/Python

// Loans row count
spark.sql("SELECT count(*) FROM loans_delta").show()

+--------+
|count(1)|
+--------+
|   14705|
+--------+

// First 5 rows of loans table
spark.sql("SELECT * FROM loans_delta LIMIT 5").show()

+-------+-----------+---------+----------+
|loan_id|funded_amnt|paid_amnt|addr_state|
+-------+-----------+---------+----------+
|      0|       1000|   182.22|        CA|
|      1|       1000|   361.19|        WA|
|      2|       1000|   176.26|        TX|
|      3|       1000|   1000.0|        OK|
|      4|       1000|   249.98|        PA|
+-------+-----------+---------+----------+
```

Loading Data Streams into a Delta Lake Table

As with static DataFrames, you can easily modify your existing Structured Streaming jobs to write to and read from a Delta Lake table by setting the format to `"delta"`. Say you have a stream of new loan data as a DataFrame named `newLoanStreamDF`, which has the same schema as the table. You can append to the table as follows:

```scala
// In Scala
import org.apache.spark.sql.streaming._

val newLoanStreamDF = ...      // Streaming DataFrame with new loans data
val checkpointDir = ...        // Directory for streaming checkpoints
val streamingQuery = newLoanStreamDF.writeStream
  .format("delta")
  .option("checkpointLocation", checkpointDir)
  .trigger(Trigger.ProcessingTime("10 seconds"))
  .start(deltaPath)
```

```python
# In Python
newLoanStreamDF = ...    # Streaming DataFrame with new loans data
checkpointDir = ...      # Directory for streaming checkpoints
streamingQuery = (newLoanStreamDF.writeStream
    .format("delta")
    .option("checkpointLocation", checkpointDir)
    .trigger(processingTime = "10 seconds")
    .start(deltaPath))
```

With this format, just like any other, Structured Streaming offers end-to-end exactly-once guarantees. However, Delta Lake has a few additional advantages over traditional formats like JSON, Parquet, or ORC:

It allows writes from both batch and streaming jobs into the same table

With other formats, data written into a table from a Structured Streaming job will overwrite any existing data in the table. This is because the metadata maintained in the table to ensure exactly-once guarantees for streaming writes does not account for other nonstreaming writes. Delta Lake's advanced metadata management allows both batch and streaming data to be written.

It allows multiple streaming jobs to append data to the same table

The same limitation of metadata with other formats also prevents multiple Structured Streaming queries from appending to the same table. Delta Lake's metadata maintains transaction information for each streaming query, thus enabling any number of streaming queries to concurrently write into a table with exactly-once guarantees.

It provides ACID guarantees even under concurrent writes

Unlike built-in formats, Delta Lake allows concurrent batch and streaming operations to write data with ACID guarantees.

Enforcing Schema on Write to Prevent Data Corruption

A common problem with managing data with Spark using common formats like JSON, Parquet, and ORC is accidental data corruption caused by writing incorrectly formatted data. Since these formats define the data layout of individual files and not of the entire table, there is no mechanism to prevent any Spark job from writing files with different schemas into existing tables. This means there are no guarantees of consistency for the entire table of many Parquet files.

The Delta Lake format records the schema as table-level metadata. Hence, all writes to a Delta Lake table can verify whether the data being written has a schema compatible with that of the table. If it is not compatible, Spark will throw an error before any data is written and committed to the table, thus preventing such accidental data corruption. Let's test this by trying to write some data with an additional column, closed, that signifies whether the loan has been terminated. Note that this column does not exist in the table:

```scala
// In Scala
val loanUpdates = Seq(
    (1111111L, 1000, 1000.0, "TX", false),
    (2222222L, 2000, 0.0, "CA", true))
  .toDF("loan_id", "funded_amnt", "paid_amnt", "addr_state", "closed")

loanUpdates.write.format("delta").mode("append").save(deltaPath)
```

```python
# In Python
from pyspark.sql.functions import *

cols = ['loan_id', 'funded_amnt', 'paid_amnt', 'addr_state', 'closed']
items = [
(1111111, 1000, 1000.0, 'TX', True),
(2222222, 2000, 0.0, 'CA', False)
]

loanUpdates = (spark.createDataFrame(items, cols)
  .withColumn("funded_amnt", col("funded_amnt").cast("int")))
loanUpdates.write.format("delta").mode("append").save(deltaPath)
```

This write will fail with the following error message:

```
org.apache.spark.sql.AnalysisException: A schema mismatch detected when writing
    to the Delta table (Table ID: 48bfa949-5a09-49ce-96cb-34090ab7d695).
To enable schema migration, please set:
'.option("mergeSchema", "true")'.

Table schema:
root
-- loan_id: long (nullable = true)
-- funded_amnt: integer (nullable = true)
-- paid_amnt: double (nullable = true)
-- addr_state: string (nullable = true)
```

```
Data schema:
root
-- loan_id: long (nullable = true)
-- funded_amnt: integer (nullable = true)
-- paid_amnt: double (nullable = true)
-- addr_state: string (nullable = true)
-- closed: boolean (nullable = true)
```

This illustrates how Delta Lake blocks writes that do not match the schema of the table. However, it also gives a hint about how to actually evolve the schema of the table using the option `mergeSchema`, as discussed next.

Evolving Schemas to Accommodate Changing Data

In our world of ever-changing data, it is possible that we might want to add this new column to the table. This new column can be explicitly added by setting the option `"mergeSchema"` to `"true"`:

```scala
// In Scala
loanUpdates.write.format("delta").mode("append")
  .option("mergeSchema", "true")
  .save(deltaPath)
```

```python
# In Python
(loanUpdates.write.format("delta").mode("append")
  .option("mergeSchema", "true")
  .save(deltaPath))
```

With this, the column `closed` will be added to the table schema, and new data will be appended. When existing rows are read, the value of the new column is considered as `NULL`. In Spark 3.0, you can also use the SQL DDL command `ALTER TABLE` to add and modify columns.

Transforming Existing Data

Delta Lake supports the DML commands `UPDATE`, `DELETE`, and `MERGE`, which allow you to build complex data pipelines. These commands can be invoked using Java, Scala, Python, and SQL, giving users the flexibility of using the commands with any APIs they are familiar with, using either DataFrames or tables. Furthermore, each of these data modification operations ensures ACID guarantees.

Let's explore this with a few examples of real-world use cases.

Updating data to fix errors

A common use case when managing data is fixing errors in the data. Suppose, upon reviewing the data, we realized that all of the loans assigned to `addr_state = 'OR'` should have been assigned to `addr_state = 'WA'`. If the loan table were a Parquet table, then to do such an update we would need to:

1. Copy all of the rows that are not affected into a new table.
2. Copy all of the rows that are affected into a DataFrame, then perform the data modification.
3. Insert the previously noted DataFrame's rows into the new table.
4. Remove the old table and rename the new table to the old table name.

In Spark 3.0, which added direct support for DML SQL operations like `UPDATE`, `DELETE`, and `MERGE`, instead of manually performing all these steps you can simply run the SQL `UPDATE` command. However, with a Delta Lake table, users can run this operation too, by using Delta Lake's programmatic APIs as follows:

```scala
// In Scala
import io.delta.tables.DeltaTable
import org.apache.spark.sql.functions._

val deltaTable = DeltaTable.forPath(spark, deltaPath)
deltaTable.update(
  col("addr_state") === "OR",
  Map("addr_state" -> lit("WA")))
```

```python
# In Python
from delta.tables import *

deltaTable = DeltaTable.forPath(spark, deltaPath)
deltaTable.update("addr_state = 'OR'", {"addr_state": "'WA'"})
```

Deleting user-related data

With data protection policies like the EU's General Data Protection Regulation (GDPR) (*https://oreil.ly/hOdBE*) coming into force, it is more important now than ever to be able to delete user data from all your tables. Say it is mandated that you have to delete the data on all loans that have been fully paid off. With Delta Lake, you can do the following:

```scala
// In Scala
val deltaTable = DeltaTable.forPath(spark, deltaPath)
deltaTable.delete("funded_amnt >= paid_amnt")
```

```python
# In Python
deltaTable = DeltaTable.forPath(spark, deltaPath)
deltaTable.delete("funded_amnt >= paid_amnt")
```

Similar to updates, with Delta Lake and Apache Spark 3.0 you can directly run the DELETE SQL command on the table.

Upserting change data to a table using merge()

A common use case is change data capture, where you have to replicate row changes made in an OLTP table to another table for OLAP workloads. To continue with our loan data example, say we have another table of new loan information, some of which are new loans and others of which are updates to existing loans. In addition, let's say this changes table has the same schema as the loan_delta table. You can upsert these changes into the table using the DeltaTable.merge() operation, which is based on the MERGE SQL command:

```scala
// In Scala
deltaTable
  .alias("t")
  .merge(loanUpdates.alias("s"), "t.loan_id = s.loan_id")
  .whenMatched.updateAll()
  .whenNotMatched.insertAll()
  .execute()
```

```python
# In Python
(deltaTable
  .alias("t")
  .merge(loanUpdates.alias("s"), "t.loan_id = s.loan_id")
  .whenMatchedUpdateAll()
  .whenNotMatchedInsertAll()
  .execute())
```

As a reminder, you can run this as a SQL MERGE command starting with Spark 3.0. Furthermore, if you have a stream of such captured changes, you can continuously apply those changes using a Structured Streaming query. The query can read the changes in micro-batches (see Chapter 8) from any streaming source, and use fore achBatch() to apply the changes in each micro-batch to the Delta Lake table.

Deduplicating data while inserting using insert-only merge

The merge operation in Delta Lake supports an extended syntax beyond that specified by the ANSI standard, including advanced features like the following:

Delete actions
  For example, MERGE ... WHEN MATCHED THEN DELETE.

Clause conditions
  For example, MERGE ... WHEN MATCHED AND <condition> THEN ....

Optional actions
  All the MATCHED and NOT MATCHED clauses are optional.

For example, UPDATE * and INSERT * to update/insert all the columns in the target table with matching columns from the source data set. The equivalent Delta Lake APIs are updateAll() and insertAll(), which we saw in the previous section.

This allows you to express many more complex use cases with little code. For example, say you want to backfill the loan_delta table with historical data on past loans. But some of the historical data may already have been inserted in the table, and you don't want to update those records because they may contain more up-to-date information. You can deduplicate by the loan_id while inserting by running the following merge operation with only the INSERT action (since the UPDATE action is optional):

```scala
// In Scala
deltaTable
  .alias("t")
  .merge(historicalUpdates.alias("s"), "t.loan_id = s.loan_id")
  .whenNotMatched.insertAll()
  .execute()
```

```python
# In Python
(deltaTable
  .alias("t")
  .merge(historicalUpdates.alias("s"), "t.loan_id = s.loan_id")
  .whenNotMatchedInsertAll()
  .execute())
```

There are even more complex use cases, like CDC with deletes and SCD tables, that are made simple with the extended merge syntax. Refer to the documentation (*https://oreil.ly/XBag7*) for more details and examples.

Auditing Data Changes with Operation History

All of the changes to your Delta Lake table are recorded as commits in the table's transaction log. As you write into a Delta Lake table or directory, every operation is automatically versioned. You can query the table's operation history as noted in the following code snippet:

```
// In Scala/Python
deltaTable.history().show()
```

By default this will show a huge table with many versions and a lot of columns. Let's instead print some of the key columns of the last three operations:

```scala
// In Scala
deltaTable
  .history(3)
  .select("version", "timestamp", "operation", "operationParameters")
  .show(false)
```

```
# In Python
(deltaTable
  .history(3)
  .select("version", "timestamp", "operation", "operationParameters")
  .show(truncate=False))
```

This will generate the following output:

```
+-------+-----------+---------+-------------------------------------------------------+
|version|timestamp  |operation|operationParameters                                    |
+-------+-----------+---------+-------------------------------------------------------+
|5      |2020-04-07 |MERGE    |[predicate -> (t.`loan_id` = s.`loan_id`)] |
|4      |2020-04-07 |MERGE    |[predicate -> (t.`loan_id` = s.`loan_id`)] |
|3      |2020-04-07 |DELETE   |[predicate -> ["(CAST(`funded_amnt` ...    |
+-------+-----------+---------+-------------------------------------------------------+
```

Note the `operation` and `operationParameters` that are useful for auditing the changes.

Querying Previous Snapshots of a Table with Time Travel

You can query previous versioned snapshots of a table by using the `DataFrameReader` options `"versionAsOf"` and `"timestampAsOf"`. Here are a few examples:

```scala
// In Scala
spark.read
  .format("delta")
  .option("timestampAsOf", "2020-01-01")  // timestamp after table creation
  .load(deltaPath)

spark.read.format("delta")
  .option("versionAsOf", "4")
  .load(deltaPath)
```

```python
# In Python
(spark.read
  .format("delta")
  .option("timestampAsOf", "2020-01-01")  # timestamp after table creation
  .load(deltaPath))

(spark.read.format("delta")
  .option("versionAsOf", "4")
  .load(deltaPath))
```

This is useful in a variety of situations, such as:

- Reproducing machine learning experiments and reports by rerunning the job on a specific table version

- Comparing the data changes between different versions for auditing

- Rolling back incorrect changes by reading a previous snapshot as a DataFrame and overwriting the table with it

---

Summary

This chapter examined the possibilities for building reliable data lakes using Apache Spark. To recap, databases have solved data problems for a long time, but they fail to fulfill the diverse requirements of modern use cases and workloads. Data lakes were built to alleviate some of the limitations of databases, and Apache Spark is one of the best tools to build them with. However, data lakes still lack some of the key features provided by databases (e.g., ACID guarantees). Lakehouses are the next generation of data solutions, which aim to provide the best features of databases and data lakes and meet all the requirements of diverse use cases and workloads.

We briefly explored a couple of open source systems (Apache Hudi and Apache Iceberg) that can be used to build lakehouses, then took a closer look at Delta Lake, a file-based open source storage format that, along with Apache Spark, is a great building block for lakehouses. As you saw, it provides the following:

- Transactional guarantees and schema management, like databases
- Scalability and openness, like data lakes
- Support for concurrent batch and streaming workloads with ACID guarantees
- Support for transformation of existing data using update, delete, and merge operations that ensure ACID guarantees
- Support for versioning, auditing of operation history, and querying of previous versions

In the next chapter, we'll explore how to begin building ML models using Spark's MLlib.

Machine Learning with MLlib

Up until this point, we have focused on data engineering workloads with Apache Spark. Data engineering is often a precursory step to preparing your data for machine learning (ML) tasks, which will be the focus of this chapter. We live in an era in which machine learning and artificial intelligence applications are an integral part of our lives. Chances are that whether we realize it or not, every day we come into contact with ML models for purposes such as online shopping recommendations and advertisements, fraud detection, classification, image recognition, pattern matching, and more. These ML models drive important business decisions for many companies. According to this McKinsey study (*https://oreil.ly/Dxj0A*), 35% of what consumers purchase on Amazon and 75% of what they watch on Netflix is driven by machine learning–based product recommendations. Building a model that performs well can make or break companies.

In this chapter we will get you started building ML models using MLlib (*https://oreil.ly/_XSOs*), the de facto machine learning library in Apache Spark. We'll begin with a brief introduction to machine learning, then cover best practices for distributed ML and feature engineering at scale (if you're already familiar with machine learning fundamentals, you can skip straight to "Designing Machine Learning Pipelines" on page 289). Through the short code snippets presented here and the notebooks available in the book's GitHub repo (*https://github.com/databricks/LearningSparkV2*), you'll learn how to build basic ML models and use MLlib.

This chapter covers the Scala and Python APIs; if you're interested in using R (sparklyr) with Spark for machine learning, we invite you to check out *Mastering Spark with R* by Javier Luraschi, Kevin Kuo, and Edgar Ruiz (O'Reilly).

What Is Machine Learning?

Machine learning is getting a lot of hype these days—but what is it, exactly? Broadly speaking, machine learning is a process for extracting patterns from your data, using statistics, linear algebra, and numerical optimization. Machine learning can be applied to problems such as predicting power consumption, determining whether or not there is a cat in your video, or clustering items with similar characteristics.

There are a few types of machine learning, including supervised, semi-supervised, unsupervised, and reinforcement learning. This chapter will mainly focus on supervised machine learning and just touch upon unsupervised learning. Before we dive in, let's briefly discuss the differences between supervised and unsupervised ML.

Supervised Learning

In supervised machine learning (*https://oreil.ly/fVOVL*), your data consists of a set of input records, each of which has associated labels, and the goal is to predict the output label(s) given a new unlabeled input. These output labels can either be *discrete* or *continuous*, which brings us to the two types of supervised machine learning: *classification* and *regression*.

In a classification problem, the aim is to separate the inputs into a discrete set of classes or labels. With *binary* classification, there are two discrete labels you want to predict, such as "dog" or "not dog," as Figure 10-1 depicts.

Figure 10-1. Binary classification example: dog or not dog

With *multiclass*, also known as *multinomial*, classification, there can be three or more discrete labels, such as predicting the breed of a dog (e.g., Australian shepherd, golden retriever, or poodle, as shown in Figure 10-2).

Figure 10-2. Multinomial classification example: Australian shepherd, golden retriever, or poodle

In regression problems, the value to predict is a continuous number, not a label. This means you might predict values that your model hasn't seen during training, as shown in Figure 10-3. For example, you might build a model to predict the daily ice cream sales given the temperature. Your model might predict the value $77.67, even if none of the input/output pairs it was trained on contained that value.

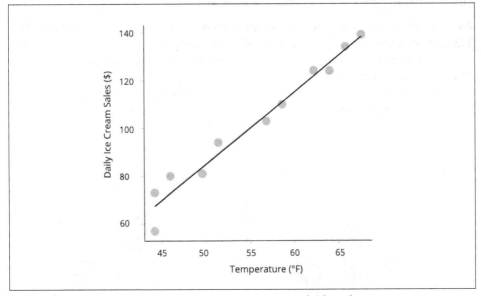

Figure 10-3. Regression example: predicting ice cream sales based on temperature

Table 10-1 lists some commonly used supervised ML algorithms that are available in Spark MLlib (*https://oreil.ly/Yt0uu*), with a note as to whether they can be used for regression, classification, or both.

Table 10-1. Popular classification and regression algorithms

Algorithm	Typical usage
Linear regression	Regression
Logistic regression	Classification (we know, it has regression in the name!)
Decision trees	Both
Gradient boosted trees	Both
Random forests	Both
Naive Bayes	Classification
Support vector machines (SVMs)	Classification

Unsupervised Learning

Obtaining the labeled data required by supervised machine learning can be very expensive and/or infeasible. This is where unsupervised machine learning (*https://oreil.ly/J80ym*) comes into play. Instead of predicting a label, unsupervised ML helps you to better understand the structure of your data.

As an example, consider the original unclustered data on the left in Figure 10-4. There is no known true label for each of these data points (x_1, x_2), but by applying unsupervised machine learning to our data we can find the clusters that naturally form, as shown on the right.

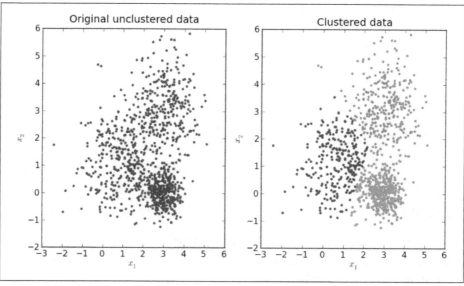

Figure 10-4. Clustering example

Unsupervised machine learning can be used for outlier detection or as a preprocessing step for supervised machine learning—for example, to reduce the dimensionality (*https://oreil.ly/N5JWF*) (i.e., number of dimensions per datum) of the data set, which is useful for reducing storage requirements or simplifying downstream tasks. Some unsupervised machine learning algorithms in MLlib (*https://oreil.ly/NLYo6*) include *k*-means, Latent Dirichlet Allocation (LDA), and Gaussian mixture models.

Why Spark for Machine Learning?

Spark is a unified analytics engine that provides an ecosystem for data ingestion, feature engineering, model training, and deployment. Without Spark, developers would need many disparate tools to accomplish this set of tasks, and might still struggle with scalability.

Spark has two machine learning packages: `spark.mllib` (*https://oreil.ly/qy-PT*) and `spark.ml` (*https://oreil.ly/WBGzN*). `spark.mllib` is the original machine learning API, based on the RDD API (which has been in maintenance mode since Spark 2.0), while `spark.ml` is the newer API, based on DataFrames. The rest of this chapter will focus on using the `spark.ml` package and how to design machine learning pipelines in Spark. However, we use "MLlib" as an umbrella term to refer to both machine learning library packages in Apache Spark.

With `spark.ml`, data scientists can use one ecosystem for their data preparation and model building, without the need to downsample their data to fit on a single machine. `spark.ml` focuses on O(*n*) scale-out, where the model scales linearly with the number of data points you have, so it can scale to massive amounts of data. In the following chapter, we will discuss some of the trade-offs involved in choosing between a distributed framework such as `spark.ml` and a single-node framework like `scikit-learn` (*https://oreil.ly/WSz_8*) (`sklearn`). If you have previously used `scikit-learn`, many of the APIs in `spark.ml` will feel quite familiar, but there are some subtle differences that we will discuss.

Designing Machine Learning Pipelines

In this section, we will cover how to create and tune ML pipelines. The concept of pipelines is common across many ML frameworks as a way to organize a series of operations to apply to your data. In MLlib, the Pipeline API (*https://oreil.ly/FdTA_*) provides a high-level API built on top of DataFrames to organize your machine learning workflow. The Pipeline API is composed of a series of transformers and estimators, which we will discuss in-depth later.

Throughout this chapter, we will use the San Francisco housing data set from Inside Airbnb (*https://oreil.ly/hBfNj*). It contains information about Airbnb rentals in San Francisco, such as the number of bedrooms, location, review scores, etc., and our goal is to build a model to predict the nightly rental prices for listings in that city. This is a regression problem, because price is a continuous variable. We will guide you through the workflow a data scientist would go through to approach this problem, including feature engineering, building models, hyperparameter tuning, and evaluating model performance. This data set is quite messy and can be difficult to model (like most real-world data sets!), so if you are experimenting on your own, don't feel bad if your early models aren't great.

The intent of this chapter is not to show you every API in MLlib, but rather to equip you with the skills and knowledge to get started with using MLlib to build end-to-end pipelines. Before going into the details, let's define some MLlib terminology:

Transformer
> Accepts a DataFrame as input, and returns a new DataFrame with one or more columns appended to it. Transformers do not learn any parameters from your data and simply apply rule-based transformations to either prepare data for model training or generate predictions using a trained MLlib model. They have a `.transform()` method.

Estimator
> Learns (or "fits") parameters from your DataFrame via a `.fit()` method and returns a `Model`, which is a transformer.

Pipeline
> Organizes a series of transformers and estimators into a single model. While pipelines themselves are estimators, the output of `pipeline.fit()` returns a `Pipe lineModel`, a transformer.

While these concepts may seem rather abstract right now, the code snippets and examples in this chapter will help you understand how they all come together. But before we can build our ML models and use transformers, estimators, and pipelines, we need to load in our data and perform some data preparation.

Data Ingestion and Exploration

We have slightly preprocessed the data in our example data set to remove outliers (e.g., Airbnbs posted for $0/night), converted all integers to doubles, and selected an informative subset of the more than one hundred fields. Further, for any missing numerical values in our data columns, we have imputed the median value and added an indicator column (the column name followed by _na, such as bedrooms_na). This way the ML model or human analyst can interpret any value in that column as an imputed value, not a true value. You can see the data preparation notebook in the book's GitHub repo (*https://github.com/databricks/LearningSparkV2*). Note there are many other ways to handle missing values, which are outside the scope of this book.

Let's take a quick peek at the data set and the corresponding schema (with the output showing just a subset of the columns):

```python
# In Python
filePath = """/databricks-datasets/learning-spark-v2/sf-airbnb/
sf-airbnb-clean.parquet/"""
airbnbDF = spark.read.parquet(filePath)
airbnbDF.select("neighbourhood_cleansed", "room_type", "bedrooms", "bathrooms",
                "number_of_reviews", "price").show(5)
```

```scala
// In Scala
val filePath =
  "/databricks-datasets/learning-spark-v2/sf-airbnb/sf-airbnb-clean.parquet/"
val airbnbDF = spark.read.parquet(filePath)
airbnbDF.select("neighbourhood_cleansed", "room_type", "bedrooms", "bathrooms",
                "number_of_reviews", "price").show(5)
```

```
+----------------------+----------------+--------+---------+----------+-----+
|neighbourhood_cleansed|       room_type|bedrooms|bathrooms|number_...|price|
+----------------------+----------------+--------+---------+----------+-----+
|      Western Addition|Entire home/apt|     1.0|      1.0|     180.0|170.0|
|        Bernal Heights|Entire home/apt|     2.0|      1.0|     111.0|235.0|
|        Haight Ashbury|    Private room|     1.0|      4.0|      17.0| 65.0|
|        Haight Ashbury|    Private room|     1.0|      4.0|       8.0| 65.0|
|      Western Addition|Entire home/apt|     2.0|      1.5|      27.0|785.0|
+----------------------+----------------+--------+---------+----------+-----+
```

Our goal is to predict the price per night for a rental property, given our features.

> Before data scientists can get to model building, they need to explore and understand their data. They will often use Spark to group their data, then use data visualization libraries such as matplotlib (*https://matplotlib.org*) to visualize the data. We will leave data exploration as an exercise for the reader.

Creating Training and Test Data Sets

Before we begin feature engineering and modeling, we will divide our data set into two groups: *train* and *test*. Depending on the size of your data set, your train/test ratio may vary, but many data scientists use 80/20 as a standard train/test split. You might be wondering, "Why not use the entire data set to train the model?" The problem is that if we built a model on the entire data set, it's possible that the model would memorize or "overfit" to the training data we provided, and we would have no more data with which to evaluate how well it generalizes to previously unseen data. The model's performance on the test set is a proxy for how well it will perform on unseen data (i.e., in the wild or in production), assuming that data follows similar distributions. This split is depicted in Figure 10-5.

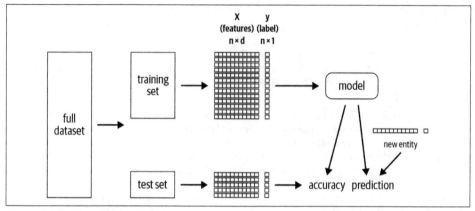

Figure 10-5. Train/test split

Our training set consists of a set of features, X, and a label, y. Here we use capital X to denote a matrix with dimensions *n* x *d*, where *n* is the number of data points (or examples) and *d* is the number of features (this is what we call the fields or columns in our DataFrame). We use lowercase y to denote a vector, with dimensions *n* x 1; for every example, there is one label.

Different metrics are used to measure the performance of the model. For classification problems, a standard metric is the *accuracy*, or percentage, of correct predictions. Once the model has satisfactory performance on the training set using that metric, we will apply the model to our test set. If it performs well on our test set according to our evaluation metrics, then we can feel confident that we have built a model that will "generalize" to unseen data.

For our Airbnb data set, we will keep 80% for the training set and set aside 20% of our data for the test set. Further, we will set a random seed for reproducibility, such that if we rerun this code we will get the same data points going to our train and test data sets, respectively. The value of the seed itself *shouldn't* matter, but data scientists often like setting it to 42 as that is the answer to the Ultimate Question of Life (*https://oreil.ly/sE12h*):

```
# In Python
trainDF, testDF = airbnbDF.randomSplit([.8, .2], seed=42)
print(f"""There are {trainDF.count()} rows in the training set,
and {testDF.count()} in the test set""")
```

```
// In Scala
val Array(trainDF, testDF) = airbnbDF.randomSplit(Array(.8, .2), seed=42)
println(f"""There are ${trainDF.count} rows in the training set, and
${testDF.count} in the test set""")
```

This produces the following output:

```
There are 5780 rows in the training set, and 1366 in the test set
```

But what happens if we change the number of executors in our Spark cluster? The Catalyst optimizer determines the optimal way to partition your data (*https://oreil.ly/ Ecd_m*) as a function of your cluster resources and size of your data set. Given that data in a Spark DataFrame is row-partitioned and each worker performs its split independently of the other workers, if the data in the partitions changes, then the result of the split (by `randomSplit()`) won't be the same.

While you could fix your cluster configuration and your seed to ensure that you get consistent results, our recommendation is to split your data once, then write it out to its own train/test folder so you don't have these reproducibility issues.

> During your exploratory analysis, you should cache the training data set because you will be accessing it many times throughout the machine learning process. Please reference the section on "Caching and Persistence of Data" on page 183 in Chapter 7.

Preparing Features with Transformers

Now that we have split our data into training and test sets, let's prepare the data to build a linear regression model predicting price given the number of bedrooms. In a later example, we will include all of the relevant features, but for now let's make sure we have the mechanics in place. Linear regression (like many other algorithms in Spark) requires that all the input features are contained within a single vector in your DataFrame. Thus, we need to *transform* our data.

Transformers in Spark accept a DataFrame as input and return a new DataFrame with one or more columns appended to it. They do not learn from your data, but apply rule-based transformations using the `transform()` method.

For the task of putting all of our features into a single vector, we will use the `VectorAssembler` transformer (*https://oreil.ly/r2MSV*). `VectorAssembler` takes a list of input columns and creates a new DataFrame with an additional column, which we will call `features`. It combines the values of those input columns into a single vector:

```python
# In Python
from pyspark.ml.feature import VectorAssembler
vecAssembler = VectorAssembler(inputCols=["bedrooms"], outputCol="features")
vecTrainDF = vecAssembler.transform(trainDF)
vecTrainDF.select("bedrooms", "features", "price").show(10)
```

```scala
// In Scala
import org.apache.spark.ml.feature.VectorAssembler
val vecAssembler = new VectorAssembler()
  .setInputCols(Array("bedrooms"))
  .setOutputCol("features")
val vecTrainDF = vecAssembler.transform(trainDF)
vecTrainDF.select("bedrooms", "features", "price").show(10)
```

```
+--------+--------+-----+
|bedrooms|features|price|
+--------+--------+-----+
|     1.0|   [1.0]|200.0|
|     1.0|   [1.0]|130.0|
|     1.0|   [1.0]| 95.0|
|     1.0|   [1.0]|250.0|
|     3.0|   [3.0]|250.0|
|     1.0|   [1.0]|115.0|
|     1.0|   [1.0]|105.0|
|     1.0|   [1.0]| 86.0|
|     1.0|   [1.0]|100.0|
|     2.0|   [2.0]|220.0|
+--------+--------+-----+
```

You'll notice that in the Scala code, we had to instantiate the new VectorAssembler object as well as using setter methods to change the input and output columns. In Python, you have the option to pass the parameters directly to the constructor of Vec torAssembler or to use the setter methods, but in Scala you can only use the setter methods.

We cover the fundamentals of linear regression next, but if you are already familiar with the algorithm, please skip to "Using Estimators to Build Models" on page 295.

Understanding Linear Regression

Linear regression (*https://oreil.ly/dhgZf*) models a linear relationship between your dependent variable (or label) and one or more independent variables (or features). In our case, we want to fit a linear regression model to predict the price of an Airbnb rental given the number of bedrooms.

In Figure 10-6, we have a single feature x and an output y (this is our dependent variable). Linear regression seeks to fit an equation for a line to x and y, which for scalar variables can be expressed as $y = mx + b$, where m is the slope and b is the offset or intercept.

The dots indicate the true (x, y) pairs from our data set, and the solid line indicates the line of best fit for this data set. The data points do not perfectly line up, so we usually think of linear regression as fitting a model to $y \approx mx + b + \varepsilon$, where ε (epsilon) is an error drawn independently per record x from some distribution. These are the errors between our model predictions and the true values. Often we think of ε as being Gaussian, or normally distributed. The vertical lines above the regression line indicate positive ε (or residuals), where your true values are above the predicted values, and the vertical lines below the regression line indicate negative residuals. The goal of linear regression is to find a line that minimizes the square of these residuals. You'll notice that the line can extrapolate predictions for data points it hasn't seen.

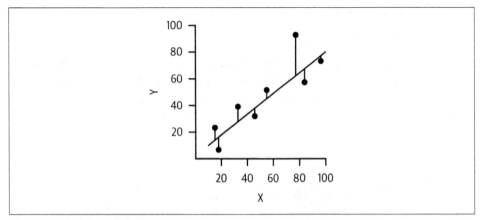

Figure 10-6. Univariate linear regression

Linear regression can also be extended to handle multiple independent variables. If we had three features as input, $x = [x_1, x_2, x_3]$, then we could model y as $y \approx w_0 + w_1 x_1 + w_2 x_2 + w_3 x_3 + \varepsilon$. In this case, there is a separate coefficient (or weight) for each feature and a single intercept (w_0 instead of b here). The process of estimating the coefficients and intercept for our model is called *learning* (or *fitting*) the parameters for the model. For right now, we'll focus on the univariate regression example of predicting price given the number of bedrooms, and we'll get back to multivariate linear regression in a bit.

Using Estimators to Build Models

After setting up our `vectorAssembler`, we have our data prepared and transformed into a format that our linear regression model expects. In Spark, `LinearRegression` (*https://oreil.ly/zxlnL*) is a type of estimator—it takes in a DataFrame and returns a `Model`. Estimators learn parameters from your data, have an `estimator_name.fit()` method, and are eagerly evaluated (i.e., kick off Spark jobs), whereas transformers are lazily evaluated. Some other examples of estimators include `Imputer`, `DecisionTree` `Classifier`, and `RandomForestRegressor`.

You'll notice that our input column for linear regression (`features`) is the output from our `vectorAssembler`:

```python
# In Python
from pyspark.ml.regression import LinearRegression
lr = LinearRegression(featuresCol="features", labelCol="price")
lrModel = lr.fit(vecTrainDF)
```

```scala
// In Scala
import org.apache.spark.ml.regression.LinearRegression
val lr = new LinearRegression()
  .setFeaturesCol("features")
```

```
        .setLabelCol("price")

    val lrModel = lr.fit(vecTrainDF)
```

`lr.fit()` returns a `LinearRegressionModel` (*https://oreil.ly/LASya*) (`lrModel`), which is a transformer. In other words, the output of an estimator's `fit()` method is a transformer. Once the estimator has learned the parameters, the transformer can apply these parameters to new data points to generate predictions. Let's inspect the parameters it learned:

```
# In Python
m = round(lrModel.coefficients[0], 2)
b = round(lrModel.intercept, 2)
print(f"""The formula for the linear regression line is
price = {m}*bedrooms + {b}""")

// In Scala
val m = lrModel.coefficients(0)
val b = lrModel.intercept
println(f"""The formula for the linear regression line is
price = $m%1.2f*bedrooms + $b%1.2f""")
```

This prints:

```
The formula for the linear regression line is price = 123.68*bedrooms + 47.51
```

Creating a Pipeline

If we want to apply our model to our test set, then we need to prepare that data in the same way as the training set (i.e., pass it through the vector assembler). Oftentimes data preparation pipelines will have multiple steps, and it becomes cumbersome to remember not only which steps to apply, but also the ordering of the steps. This is the motivation for the Pipeline API (*https://oreil.ly/MG3YM*): you simply specify the stages you want your data to pass through, in order, and Spark takes care of the processing for you. They provide the user with better code reusability and organization. In Spark, `Pipelines` are estimators, whereas `PipelineModels`—fitted `Pipelines`—are transformers.

Let's build our pipeline now:

```
# In Python
from pyspark.ml import Pipeline
pipeline = Pipeline(stages=[vecAssembler, lr])
pipelineModel = pipeline.fit(trainDF)

// In Scala
import org.apache.spark.ml.Pipeline
val pipeline = new Pipeline().setStages(Array(vecAssembler, lr))
val pipelineModel = pipeline.fit(trainDF)
```

Another advantage of using the Pipeline API is that it determines which stages are estimators/transformers for you, so you don't have to worry about specifying *name*.fit() versus *name*.transform() for each of the stages.

Since pipelineModel is a transformer, it is straightforward to apply it to our test data set too:

```python
# In Python
predDF = pipelineModel.transform(testDF)
predDF.select("bedrooms", "features", "price", "prediction").show(10)
```

```scala
// In Scala
val predDF = pipelineModel.transform(testDF)
predDF.select("bedrooms", "features", "price", "prediction").show(10)
```

```
+--------+--------+------+------------------+
|bedrooms|features| price|        prediction|
+--------+--------+------+------------------+
|     1.0|   [1.0]|  85.0|171.18598011578285|
|     1.0|   [1.0]|  45.0|171.18598011578285|
|     1.0|   [1.0]|  70.0|171.18598011578285|
|     1.0|   [1.0]| 128.0|171.18598011578285|
|     1.0|   [1.0]| 159.0|171.18598011578285|
|     2.0|   [2.0]| 250.0|294.86172649777757|
|     1.0|   [1.0]|  99.0|171.18598011578285|
|     1.0|   [1.0]|  95.0|171.18598011578285|
|     1.0|   [1.0]| 100.0|171.18598011578285|
|     1.0|   [1.0]|2010.0|171.18598011578285|
+--------+--------+------+------------------+
```

In this code we built a model using only a single feature, bedrooms (you can find the notebook for this chapter in the book's GitHub repo (*https://github.com/databricks/ LearningSparkV2*)). However, you may want to build a model using all of your features, some of which may be categorical, such as host_is_superhost. Categorical features take on discrete values and have no intrinsic ordering—examples include occupations or country names. In the next section we'll consider a solution for how to treat these kinds of variables, known as *one-hot encoding*.

One-hot encoding

In the pipeline we just created, we only had two stages, and our linear regression model only used one feature. Let's take a look at how to build a slightly more complex pipeline that incorporates all of our numeric and categorical features.

Most machine learning models in MLlib expect numerical values as input, represented as vectors. To convert categorical values into numeric values, we can use a technique called one-hot encoding (OHE). Suppose we have a column called Animal and we have three types of animals: Dog, Cat, and Fish. We can't pass the string types into our ML model directly, so we need to assign a numeric mapping, such as this:

```
Animal = {"Dog", "Cat", "Fish"}
"Dog" = 1, "Cat" = 2, "Fish" = 3
```

However, using this approach we've introduced some spurious relationships into our data set that weren't there before. For example, why did we assign Cat twice the value of Dog? The numeric values we use should not introduce any relationships into our data set. Instead, we want to create a separate column for every distinct value in our Animal column:

```
"Dog" = [ 1, 0, 0]
"Cat" = [ 0, 1, 0]
"Fish" = [0, 0, 1]
```

If the animal is a dog, it has a one in the first column and zeros elsewhere. If it is a cat, it has a one in the second column and zeros elsewhere. The ordering of the columns is irrelevant. If you've used pandas before, you'll note that this does the same thing as pandas.get_dummies() (*https://oreil.ly/4BsUq*).

If we had a zoo of 300 animals, would OHE massively increase consumption of memory/compute resources? Not with Spark! Spark internally uses a SparseVector (*https://oreil.ly/7rOcC*) when the majority of the entries are 0, as is often the case after OHE, so it does not waste space storing 0 values. Let's take a look at an example to better understand how SparseVectors work:

```
DenseVector(0, 0, 0, 7, 0, 2, 0, 0, 0, 0)
SparseVector(10, [3, 5], [7, 2])
```

The DenseVector (*https://oreil.ly/F37Ht*) in this example contains 10 values, all but 2 of which are 0. To create a SparseVector, we need to keep track of the size of the vector, the indices of the nonzero elements, and the corresponding values at those indices. In this example the size of the vector is 10, there are two nonzero values at indices 3 and 5, and the corresponding values at those indices are 7 and 2.

There are a few ways to one-hot encode your data with Spark. A common approach is to use the StringIndexer (*https://oreil.ly/mqGP6*) and OneHotEncoder (*https://oreil.ly/D07R0*). With this approach, the first step is to apply the StringIndexer estimator to convert categorical values into category indices. These category indices are ordered by label frequencies, so the most frequent label gets index 0, which provides us with reproducible results across various runs of the same data.

Once you have created your category indices, you can pass those as input to the OneHotEncoder (OneHotEncoderEstimator (*https://oreil.ly/SmZTw*) if using Spark 2.3/2.4). The OneHotEncoder maps a column of category indices to a column of binary vectors. Take a look at Table 10-2 to see the differences in the StringIndexer and OneHotEncoder APIs from Spark 2.3/2.4 to 3.0.

Table 10-2. StringIndexer and OneHotEncoder changes in Spark 3.0

	Spark 2.3 and 2.4	Spark 3.0
StringIndexer	Single column as input/output	Multiple columns as input/output
OneHotEncoder	Deprecated	Multiple columns as input/output
OneHotEncoderEstimator	Multiple columns as input/output	N/A

The following code demonstrates how to one-hot encode our categorical features. In our data set, any column of type `string` is treated as a categorical feature, but sometimes you might have numeric features you want treated as categorical or vice versa. You'll need to carefully identify which columns are numeric and which are categorical:

```python
# In Python
from pyspark.ml.feature import OneHotEncoder, StringIndexer

categoricalCols = [field for (field, dataType) in trainDF.dtypes
                   if dataType == "string"]
indexOutputCols = [x + "Index" for x in categoricalCols]
oheOutputCols = [x + "OHE" for x in categoricalCols]

stringIndexer = StringIndexer(inputCols=categoricalCols,
                              outputCols=indexOutputCols,
                              handleInvalid="skip")
oheEncoder = OneHotEncoder(inputCols=indexOutputCols,
                           outputCols=oheOutputCols)

numericCols = [field for (field, dataType) in trainDF.dtypes
               if ((dataType == "double") & (field != "price"))]
assemblerInputs = oheOutputCols + numericCols
vecAssembler = VectorAssembler(inputCols=assemblerInputs,
                               outputCol="features")
```

```scala
// In Scala
import org.apache.spark.ml.feature.{OneHotEncoder, StringIndexer}

val categoricalCols = trainDF.dtypes.filter(_._2 == "StringType").map(_._1)
val indexOutputCols = categoricalCols.map(_ + "Index")
val oheOutputCols = categoricalCols.map(_ + "OHE")

val stringIndexer = new StringIndexer()
  .setInputCols(categoricalCols)
  .setOutputCols(indexOutputCols)
  .setHandleInvalid("skip")

val oheEncoder = new OneHotEncoder()
  .setInputCols(indexOutputCols)
  .setOutputCols(oheOutputCols)

val numericCols = trainDF.dtypes.filter{ case (field, dataType) =>
  dataType == "DoubleType" && field != "price"}.map(_._1)
```

```
val assemblerInputs = oheOutputCols ++ numericCols
val vecAssembler = new VectorAssembler()
  .setInputCols(assemblerInputs)
  .setOutputCol("features")
```

Now you might be wondering, "How does the StringIndexer handle new categories that appear in the test data set, but not in the training data set?" There is a handleInvalid parameter that specifies how you want to handle them. The options are skip (filter out rows with invalid data), error (throw an error), or keep (put invalid data in a special additional bucket, at index numLabels). For this example, we just skipped the invalid records.

One difficulty with this approach is that you need to tell StringIndexer explicitly which features should be treated as categorical features. You could use VectorIndexer (*https://oreil.ly/tNE1d*) to automatically detect all the categorical variables, but it is computationally expensive as it has to iterate over every single column and detect if it has fewer than maxCategories distinct values. maxCategories is a parameter the user specifies, and determining this value can also be difficult.

Another approach is to use RFormula (*https://oreil.ly/Jh7Q9*). The syntax for this is inspired by the R programming language. With RFormula, you provide your label and which features you want to include. It supports a limited subset of the R operators, including ~, ., :, +, and -. For example, you might specify formula = "y ~ bedrooms + bathrooms", which means to predict y given just bedrooms and bathrooms, or formula = "y ~ .", which means to use all of the available features (and automatically excludes y from the features). RFormula will automatically StringIndex and OHE all of your string columns, convert your numeric columns to double type, and combine all of these into a single vector using VectorAssembler under the hood. Thus, we can replace all of the preceding code with a single line, and we will get the same result:

```python
# In Python
from pyspark.ml.feature import RFormula

rFormula = RFormula(formula="price ~ .",
                    featuresCol="features",
                    labelCol="price",
                    handleInvalid="skip")
```

```scala
// In Scala
import org.apache.spark.ml.feature.RFormula

val rFormula = new RFormula()
  .setFormula("price ~ .")
  .setFeaturesCol("features")
  .setLabelCol("price")
  .setHandleInvalid("skip")
```

The downside of RFormula automatically combining the StringIndexer and OneHotEncoder is that one-hot encoding is not required or recommended for all algorithms. For example, tree-based algorithms can handle categorical variables directly if you just use the StringIndexer for the categorical features. You do not need to one-hot encode categorical features for tree-based methods, and it will often make your tree-based models worse (*https://oreil.ly/xfR-_*). Unfortunately, there is no one-size-fits-all solution for feature engineering, and the ideal approach is closely related to the downstream algorithms you plan to apply to your data set.

> If someone else performs the feature engineering for you, make sure they document how they generated those features.

Once you've written the code to transform your data set, you can add a linear regression model using all of the features as input.

Here, we put all the feature preparation and model building into the pipeline, and apply it to our data set:

```python
# In Python
lr = LinearRegression(labelCol="price", featuresCol="features")
pipeline = Pipeline(stages = [stringIndexer, oheEncoder, vecAssembler, lr])
# Or use RFormula
# pipeline = Pipeline(stages = [rFormula, lr])

pipelineModel = pipeline.fit(trainDF)
predDF = pipelineModel.transform(testDF)
predDF.select("features", "price", "prediction").show(5)
```

```scala
// In Scala
val lr = new LinearRegression()
  .setLabelCol("price")
  .setFeaturesCol("features")
val pipeline = new Pipeline()
  .setStages(Array(stringIndexer, oheEncoder, vecAssembler, lr))
// Or use RFormula
// val pipeline = new Pipeline().setStages(Array(rFormula, lr))

val pipelineModel = pipeline.fit(trainDF)
val predDF = pipelineModel.transform(testDF)
predDF.select("features", "price", "prediction").show(5)
```

```
+--------------------+-----+------------------+
|            features|price|        prediction|
+--------------------+-----+------------------+
|(98,[0,3,6,7,23,4...| 85.0| 55.80250714362137|
|(98,[0,3,6,7,23,4...| 45.0| 22.74720286761658|
|(98,[0,3,6,7,23,4...| 70.0|27.115811183814913|
```

```
|(98,[0,3,6,7,13,4...|128.0|-91.60763412465076|
|(98,[0,3,6,7,13,4...|159.0| 94.70374072351933|
+-------------------+-----+------------------+
```

As you can see, the features column is represented as a `SparseVector`. There are 98 features after one-hot encoding, followed by the nonzero indices and then the values themselves. You can see the whole output if you pass in `truncate=False` to `show()`.

How is our model performing? You can see that while some of the predictions might be considered "close," others are far off (a negative price for a rental!?). Next, we'll numerically evaluate how well our model performs across our entire test set.

Evaluating Models

Now that we have built a model, we need to evaluate how well it performs. In `spark.ml` there are classification, regression, clustering, and ranking evaluators (introduced in Spark 3.0). Given that this is a regression problem, we will use root-mean-square error (RMSE) (*https://oreil.ly/mAQXq*) and R^2 (*https://oreil.ly/nE8Cp*) (pronounced "R-squared") to evaluate our model's performance.

RMSE

RMSE is a metric that ranges from zero to infinity. The closer it is to zero, the better.

Let's walk through the mathematical formula step by step:

1. Compute the difference (or error) between the true value y_i and the predicted value $\hat{y}_i$ (pronounced y-hat, where the "hat" indicates that it is a predicted value of the quantity under the hat):

   $$\text{Error} = (y_i - \hat{y}_i)$$

2. Square the difference between y_i and $\hat{y}_i$ so that our positive and negative residuals do not cancel out. This is known as the squared error:

   $$\text{Square Error (SE)} = (y_i - \hat{y}_i)^2$$

3. Then we sum up the squared error for all n of our data points, known as the sum of squared errors (SSE) or sum of squared residuals:

   $$\text{Sum of Squared Errors (SSE)} = \sum_{i=1}^{n} (y_i - \hat{y}_i)^2$$

4. However, the SSE grows with the number of records n in the data set, so we want to normalize it by the number of records. The gives us the mean-squared error (MSE), a very commonly used regression metric:

$$\text{Mean Squared Error (MSE)} = \frac{1}{n} \sum_{i=1}^{n} (y_i - \hat{y}_i)^2$$

5. If we stop at MSE, then our error term is on the scale of $unit^2$. We'll often take the square root of the MSE to get the error back on the scale of the original unit, which gives us the root-mean-square error (RMSE):

$$\text{Root Mean Squared Error (RMSE)} = \sqrt{\frac{1}{n} \sum_{i=1}^{n} (y_i - \hat{y}_i)^2}$$

Let's evaluate our model using RMSE:

```python
# In Python
from pyspark.ml.evaluation import RegressionEvaluator
regressionEvaluator = RegressionEvaluator(
  predictionCol="prediction",
  labelCol="price",
  metricName="rmse")
rmse = regressionEvaluator.evaluate(predDF)
print(f"RMSE is {rmse:.1f}")
```

```scala
// In Scala
import org.apache.spark.ml.evaluation.RegressionEvaluator
val regressionEvaluator = new RegressionEvaluator()
  .setPredictionCol("prediction")
  .setLabelCol("price")
  .setMetricName("rmse")
val rmse = regressionEvaluator.evaluate(predDF)
println(f"RMSE is $rmse%.1f")
```

This produces the following output:

```
RMSE is 220.6
```

Interpreting the value of RMSE. So how do we know if 220.6 is a good value for the RMSE? There are various ways to interpret this value, one of which is to build a simple baseline model and compute its RMSE to compare against. A common baseline model for regression tasks is to compute the average value of the label on the training set $\bar{y}$ (pronounced y-bar), then predict $\bar{y}$ for every record in the test data set and compute the resulting RMSE (example code is available in the book's GitHub repo (*https://github.com/databricks/LearningSparkV2*)). If you try this, you will see that our baseline model has an RMSE of 240.7, so we beat our baseline. If you don't beat the baseline, then something probably went wrong in your model building process.

 If this were a classification problem, you might want to predict the most prevalent class as your baseline model.

Keep in mind that the unit of your label directly impacts your RMSE. For example, if your label is height, then your RMSE will be higher if you use centimeters rather than meters as your unit of measurement. You could arbitrarily decrease the RMSE by using a different unit, which is why it is important to compare your RMSE against a baseline.

There are also some metrics that naturally give you an intuition of how you are performing against a baseline, such as R^2, which we discuss next.

R^2

Despite the name R^2 containing "squared," R^2 values range from negative infinity to 1. Let's take a look at the math behind this metric. R^2 is computed as follows:

$$R^2 = 1 - \frac{SS_{res}}{SS_{tot}}$$

where SS_{tot} is the total sum of squares if you always predict $\bar{y}$:

$$SS_{tot} = \sum_{i=1}^{n} (y_i - \bar{y})^2$$

and SS_{res} is the sum of residuals squared from your model predictions (also known as the sum of squared errors, which we used to compute the RMSE):

$$SS_{res} = \sum_{i=1}^{n} (y_i - \hat{y}_i)^2$$

If your model perfectly predicts every data point, then your SS_{res} = 0, making your R^2 = 1. And if your SS_{res} = SS_{tot}, then the fraction is 1/1, so your R^2 is 0. This is what happens if your model performs the same as always predicting the average value, $\bar{y}$.

But what if your model performs worse than always predicting $\bar{y}$ and your SS_{res} is really large? Then your R^2 can actually be negative! If your R^2 is negative, you should reevaluate your modeling process. The nice thing about using R^2 is that you don't necessarily need to define a baseline model to compare against.

If we want to change our regression evaluator to use R^2, instead of redefining the regression evaluator, we can set the metric name using the setter property:

```python
# In Python
r2 = regressionEvaluator.setMetricName("r2").evaluate(predDF)
print(f"R2 is {r2}")
```

```scala
// In Scala
val r2 = regressionEvaluator.setMetricName("r2").evaluate(predDF)
println(s"R2 is $r2")
```

The output is:

```
R2 is 0.159854
```

Our R^2 is positive, but it's very close to 0. One of the reasons why our model is not performing too well is because our label, price, appears to be log-normally distributed (*https://oreil.ly/0Patq*). If a distribution is log-normal, it means that if we take the logarithm of the value, the result looks like a normal distribution. Price is often log-normally distributed. If you think about rental prices in San Francisco, most cost around $200 per night, but there are some that rent for thousands of dollars a night! You can see the distribution of our Airbnb prices for our training Dataset in Figure 10-7.

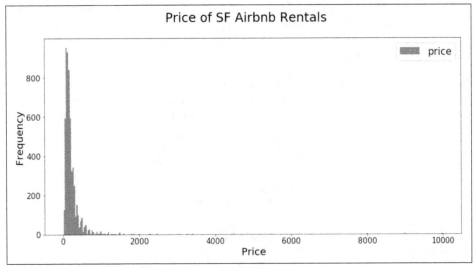

Figure 10-7. San Francisco housing price distribution

Let's take a look at the resulting distribution if we instead look at the log of the price (Figure 10-8).

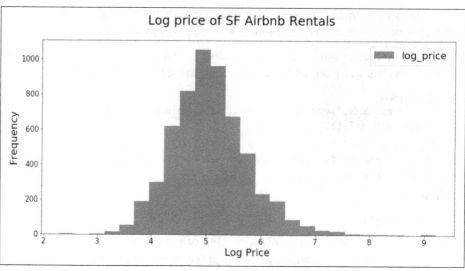

Figure 10-8. San Francisco housing log-price distribution

You can see here that our log-price distribution looks a bit more like a normal distribution. As an exercise, try building a model to predict price on the log scale, then exponentiate the prediction to get it out of log scale and evaluate your model. The code can also be found in this chapter's notebook in the book's GitHub repo (*https://github.com/databricks/LearningSparkV2*). You should see that your RMSE decreases and your R^2 increases for this data set.

Saving and Loading Models

Now that we have built and evaluated a model, let's save it to persistent storage for reuse later (or in the event that our cluster goes down, we don't have to recompute the model). Saving models is very similar to writing DataFrames—the API is `model.write().save(path)`. You can optionally provide the `overwrite()` command to overwrite any data contained in that path:

```
# In Python
pipelinePath = "/tmp/lr-pipeline-model"
pipelineModel.write().overwrite().save(pipelinePath)

// In Scala
val pipelinePath = "/tmp/lr-pipeline-model"
pipelineModel.write.overwrite().save(pipelinePath)
```

When you load your saved models, you need to specify the type of model you are loading back in (e.g., was it a `LinearRegressionModel` or a `LogisticRegressionModel`?). For this reason, we recommend you always put your transformers/estimators into a `Pipeline`, so that for all your models you load a `PipelineModel` and only need to change the file path to the model:

```
# In Python
from pyspark.ml import PipelineModel
savedPipelineModel = PipelineModel.load(pipelinePath)

// In Scala
import org.apache.spark.ml.PipelineModel
val savedPipelineModel = PipelineModel.load(pipelinePath)
```

After loading, you can apply it to new data points. However, you can't use the weights from this model as initialization parameters for training a new model (as opposed to starting with random weights), as Spark has no concept of "warm starts." If your data set changes slightly, you'll have to retrain the entire linear regression model from scratch.

With our linear regression model built and evaluated, let's explore how a few other models perform on our data set. In the next section, we will explore tree-based models and look at some common hyperparameters to tune in order to improve model performance.

Hyperparameter Tuning

When data scientists talk about tuning their models, they often discuss tuning hyperparameters to improve the model's predictive power. A *hyperparameter* is an attribute that you define about the model prior to training, and it is not learned during the training process (not to be confused with parameters, which *are* learned in the training process). The number of trees in your random forest is an example of a hyperparameter.

In this section, we will focus on using tree-based models as an example for hyperparameter tuning procedures, but the same concepts apply to other models as well. Once we set up the mechanics to do hyperparameter tuning with spark.ml, we will discuss ways to optimize the pipeline. Let's get started with a brief introduction to decision trees, followed by how we can use them in spark.ml.

Tree-Based Models

Tree-based models such as decision trees, gradient boosted trees, and random forests are relatively simple yet powerful models that are easy to interpret (meaning, it is easy to explain the predictions they make). Hence, they're quite popular for machine learning tasks. We'll get to random forests shortly, but first we need to cover the fundamentals of decision trees.

Decision trees

As an off-the-shelf solution, decision trees are well suited to data mining. They are relatively fast to build, highly interpretable, and scale-invariant (i.e., standardizing or scaling the numeric features does not change the performance of the tree). So what is a decision tree?

A decision tree is a series of if-then-else rules learned from your data for classification or regression tasks. Suppose we are trying to build a model to predict whether or not someone will accept a job offer, and the features comprise salary, commute time, free coffee, etc. If we fit a decision tree to this data set, we might get a model that looks like Figure 10-9.

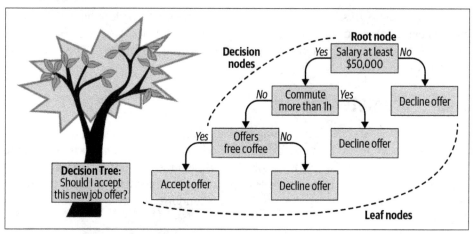

Figure 10-9. Decision tree example

The node at the top of the tree is called the "root" of the tree because it's the first feature that we "split" on. This feature should give the most informative split—in this case, if the salary is less than $50,000, then the majority of candidates will decline the job offer. The "Decline offer" node is known as a "leaf node" as there are no other splits coming out of that node; it's at the end of a branch. (Yes, it's a bit funny that we call it a decision "tree" but draw the root of the tree at the top and the leaves at the bottom!)

However, if the salary offered is greater than $50,000, we proceed to the next most informative feature in the decision tree, which in this case is the commute time. Even if the salary is over $50,000, if the commute is longer than one hour, then the majority of people will decline the job offer.

We won't get into the details of how to determine which features will give you the highest information gain here, but if you're interested, check out Chapter 9 of *The Elements of Statistical Learning* (*https://oreil.ly/VHVOW*), by Trevor Hastie, Robert Tibshirani, and Jerome Friedman (Springer).

The final feature in our model is free coffee. In this case the decision tree shows that if the salary is greater than $50,000, the commute is less than an hour, and there is free coffee, then the majority of people will accept our job offer (if only it were that simple!). As a follow-up resource, R2D3 (*https://oreil.ly/uKD8q*) has a great visualization of how decision trees work.

It is possible to split on the same feature multiple times in a single decision tree, but each split will occur at a different value.

The *depth* of a decision tree is the longest path from the root node to any given leaf node. In Figure 10-9, the depth is three. Trees that are very deep are prone to overfitting, or memorizing noise in your training data set, but trees that are too shallow will underfit to your data set (i.e., could have picked up more signal from the data).

With the essence of a decision tree explained, let's resume the topic of feature preparation for decision trees. For decision trees, you don't have to worry about standardizing or scaling your input features, because this has no impact on the splits—but you do have to be careful about how you prepare your categorical features.

Tree-based methods can naturally handle categorical variables. In `spark.ml`, you just need to pass the categorical columns to the `StringIndexer`, and the decision tree can take care of the rest. Let's fit a decision tree to our data set:

```python
# In Python
from pyspark.ml.regression import DecisionTreeRegressor

dt = DecisionTreeRegressor(labelCol="price")

# Filter for just numeric columns (and exclude price, our label)
numericCols = [field for (field, dataType) in trainDF.dtypes
               if ((dataType == "double") & (field != "price"))]

# Combine output of StringIndexer defined above and numeric columns
assemblerInputs = indexOutputCols + numericCols
vecAssembler = VectorAssembler(inputCols=assemblerInputs, outputCol="features")

# Combine stages into pipeline
stages = [stringIndexer, vecAssembler, dt]
```

```
pipeline = Pipeline(stages=stages)
pipelineModel = pipeline.fit(trainDF) # This line should error

// In Scala
import org.apache.spark.ml.regression.DecisionTreeRegressor

val dt = new DecisionTreeRegressor()
  .setLabelCol("price")

// Filter for just numeric columns (and exclude price, our label)
val numericCols = trainDF.dtypes.filter{ case (field, dataType) =>
  dataType == "DoubleType" && field != "price"}.map(_._1)

// Combine output of StringIndexer defined above and numeric columns
val assemblerInputs = indexOutputCols ++ numericCols
val vecAssembler = new VectorAssembler()
  .setInputCols(assemblerInputs)
  .setOutputCol("features")

// Combine stages into pipeline
val stages = Array(stringIndexer, vecAssembler, dt)
val pipeline = new Pipeline()
  .setStages(stages)

val pipelineModel = pipeline.fit(trainDF) // This line should error
```

This produces the following error:

```
java.lang.IllegalArgumentException: requirement failed: DecisionTree requires
maxBins (= 32) to be at least as large as the number of values in each
categorical feature, but categorical feature 3 has 36 values. Consider removing
this and other categorical features with a large number of values, or add more
training examples.
```

We can see that there is an issue with the maxBins parameter. What does that parameter do? maxBins determines the number of bins into which your continuous features are discretized, or split. This discretization step is crucial for performing distributed training. There is no maxBins parameter in scikit-learn because all of the data and the model reside on a single machine. In Spark, however, workers have all the columns of the data, but only a subset of the rows. Thus, when communicating about which features and values to split on, we need to be sure they're all talking about the same split values, which we get from the common discretization set up at training time. Let's take a look at Figure 10-10, which shows the PLANET (*https://oreil.ly/a0teT*) implementation of distributed decision trees, to get a better understanding of distributed machine learning and illustrate the maxBins parameter.

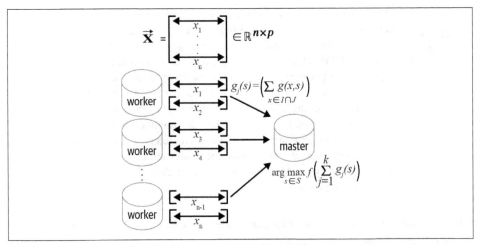

Figure 10-10. PLANET implementation of distributed decision trees (source: https://oreil.ly/RAvvP)

Every worker has to compute summary statistics for every feature and every possible split point, and those statistics will be aggregated across the workers. MLlib requires `maxBins` to be large enough to handle the discretization of the categorical columns. The default value for `maxBins` is 32, and we had a categorical column with 36 distinct values, which is why we got the error earlier. While we could increase `maxBins` to 64 to more accurately represent our continuous features, that would double the number of possible splits for continuous variables, greatly increasing our computation time. Let's instead set `maxBins` to be 40 and retrain the pipeline. You'll notice here that we are using the setter method `setMaxBins()` to modify the decision tree rather than redefining it completely:

```python
# In Python
dt.setMaxBins(40)
pipelineModel = pipeline.fit(trainDF)
```

```scala
// In Scala
dt.setMaxBins(40)
val pipelineModel = pipeline.fit(trainDF)
```

Due to differences in implementation, oftentimes you won't get exactly the same results when building a model with scikit-learn versus MLlib. However, that's OK. The key is to understand why they are different, and to see what parameters are in your control to get them to perform the way you need them to. If you are porting workloads over from scikit-learn to MLlib, we encourage you to take a look at the spark.ml (*https://oreil.ly/qFgc5*) and scikit-learn (*https://scikit-learn.org/stable*) documentation to see what parameters differ, and to tweak those parameters to get comparable results for the same data. Once the values are close enough, you can scale up your MLlib model to larger data sizes that scikit-learn can't handle.

Now that we have successfully built our model, we can extract the if-then-else rules learned by the decision tree:

```
# In Python
dtModel = pipelineModel.stages[-1]
print(dtModel.toDebugString)
```

```
// In Scala
val dtModel = pipelineModel.stages.last
  .asInstanceOf[org.apache.spark.ml.regression.DecisionTreeRegressionModel]
println(dtModel.toDebugString)
```

```
DecisionTreeRegressionModel: uid=dtr_005040f1efac, depth=5, numNodes=47,...
  If (feature 12 <= 2.5)
   If (feature 12 <= 1.5)
    If (feature 5 in {1.0,2.0})
     If (feature 4 in {0.0,1.0,3.0,5.0,9.0,10.0,11.0,13.0,14.0,16.0,18.0,24.0})
      If (feature 3 in
{0.0,1.0,2.0,3.0,4.0,5.0,6.0,7.0,8.0,9.0,10.0,11.0,12.0,13.0,14.0,...})
       Predict: 104.23992784125075
      Else (feature 3 not in {0.0,1.0,2.0,3.0,4.0,5.0,6.0,7.0,8.0,9.0,10.0,...})
       Predict: 250.7111111111111
  ...
```

This is just a subset of the printout, but you'll notice that it's possible to split on the same feature more than once (e.g., feature 12), but at different split values. Also notice the difference between how the decision tree splits on numeric features versus categorical features: for numeric features it checks if the value is less than or equal to the threshold, and for categorical features it checks if the value is in that set or not.

We can also extract the feature importance scores from our model to see the most important features:

```
# In Python
import pandas as pd

featureImp = pd.DataFrame(
```

```
  list(zip(vecAssembler.getInputCols(), dtModel.featureImportances)),
    columns=["feature", "importance"])
featureImp.sort_values(by="importance", ascending=False)

// In Scala
val featureImp = vecAssembler
  .getInputCols.zip(dtModel.featureImportances.toArray)
val columns = Array("feature", "Importance")
val featureImpDF = spark.createDataFrame(featureImp).toDF(columns: _*)

featureImpDF.orderBy($"Importance".desc).show()
```

Feature	Importance
bedrooms	0.283406
cancellation_policyIndex	0.167893
instant_bookableIndex	0.140081
property_typeIndex	0.128179
number_of_reviews	0.126233
neighbourhood_cleansedIndex	0.056200
longitude	0.038810
minimum_nights	0.029473
beds	0.015218
room_typeIndex	0.010905
accommodates	0.003603

While decision trees are very flexible and easy to use, they are not always the most accurate model. If we were to compute our R^2 on the test data set, we would actually get a negative score! That's worse than just predicting the average. (You can see this in this chapter's notebook in the book's GitHub repo (*https://github.com/databricks/Lear ningSparkV2*).)

Let's look at improving this model by using an *ensemble* approach that combines different models to achieve a better result: random forests.

Random forests

Ensembles (*https://oreil.ly/DoQPU*) work by taking a democratic approach. Imagine there are many M&Ms in a jar. You ask one hundred people to guess the number of M&Ms, and then take the average of all the guesses. The average is probably closer to the true value than most of the individual guesses. That same concept applies to machine learning models. If you build many models and combine/average their predictions, they will be more robust than those produced by any individual model.

Random forests (*https://oreil.ly/kpfTc*) are an ensemble of decision trees with two key tweaks:

Bootstrapping samples by rows

Bootstrapping is a technique for simulating new data by sampling with replacement from your original data. Each decision tree is trained on a different bootstrap sample of your data set, which produces slightly different decision trees, and then you aggregate their predictions. This technique is known as *bootstrap aggregating* (*https://oreil.ly/CfWIe*), or *bagging*. In a typical random forest implementation, each tree samples the same number of data points with replacement from the original data set, and that number can be controlled through the subsam plingRate parameter.

Random feature selection by columns

The main drawback with bagging is that the trees are all highly correlated, and thus learn similar patterns in your data. To mitigate this problem, each time you want to make a split you only consider a random subset of the columns (1/3 of the features for RandomForestRegressor and $\sqrt{\text{\#features}}$ for RandomForestClas sifier). Due to this randomness you introduce, you typically want each tree to be quite shallow. You might be thinking: each of these trees will perform worse than any single decision tree, so how could this approach possibly be better? It turns out that each of the trees learns something different about your data set, and combining this collection of "weak" learners into an ensemble makes the forest much more robust than a single decision tree.

Figure 10-11 illustrates a random forest at training time. At each split, it considers 3 of the 10 original features to split on; finally, it picks the best from among those.

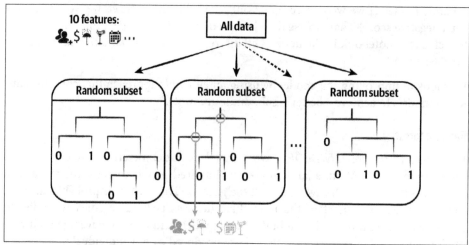

Figure 10-11. Random forest training

The APIs for random forests and decision trees are similar, and both can be applied to regression or classification tasks:

```python
# In Python
from pyspark.ml.regression import RandomForestRegressor
rf = RandomForestRegressor(labelCol="price", maxBins=40, seed=42)
```

```scala
// In Scala
import org.apache.spark.ml.regression.RandomForestRegressor
val rf = new RandomForestRegressor()
  .setLabelCol("price")
  .setMaxBins(40)
  .setSeed(42)
```

Once you've trained your random forest, you can pass new data points through the different trees trained in the ensemble.

As Figure 10-12 shows, if you build a random forest for classification, it passes the test point through each of the trees in the forest and takes a majority vote among the predictions of the individual trees. (By contrast, in regression, the random forest simply averages those predictions.) Even though each of these trees is less performant than any individual decision tree, the collection (or ensemble) actually provides a more robust model.

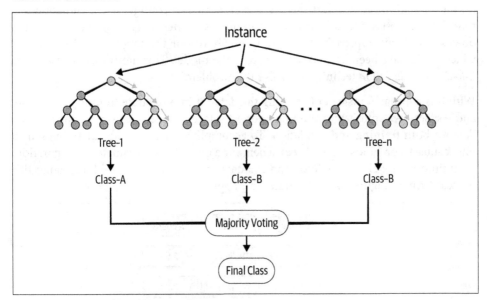

Figure 10-12. Random forest predictions

Random forests truly demonstrate the power of distributed machine learning with Spark, as each tree can be built independently of the other trees (e.g., you do not need to build tree 3 before you build tree 10). Furthermore, within each level of the tree, you can parallelize the work to find the optimal splits.

So how do we determine what the optimal number of trees in our random forest or the max depth of those trees should be? This process is called *hyperparameter tuning* (*https://oreil.ly/SqPGA*). In contrast to a parameter, a hyperparameter is a value that controls the learning process or structure of your model, and it is not learned during training. Both the number of trees and the max depth are examples of hyperparameters you can tune for random forests. Let's now shift our focus to how we can discover and evaluate the best random forest model by tuning some hyperparameters.

k-Fold Cross-Validation

Which data set should we use to determine the optimal hyperparameter values? If we use the training set, then the model is likely to overfit, or memorize the nuances of our training data. This means it will be less likely to generalize to unseen data. But if we use the test set, then that will no longer represent "unseen" data, so we won't be able to use it to verify how well our model generalizes. Thus, we need another data set to help us determine the optimal hyperparameters: the *validation* data set.

For example, instead of splitting our data into an 80/20 train/test split, as we did earlier, we can do a 60/20/20 split to generate training, validation, and test data sets, respectively. We can then build our model on the training set, evaluate performance on the validation set to select the best hyperparameter configuration, and apply the model to the test set to see how well it performs on new data. However, one of the downsides of this approach is that we lose 25% of our training data (80% -> 60%), which could have been used to help improve the model. This motivates the use of the *k-fold cross-validation* technique to solve this problem.

With this approach, instead of splitting the data set into separate training, validation, and test sets, we split it into training and test sets as before—but we use the training data for both training and validation. To accomplish this, we split our training data into *k* subsets, or "folds" (e.g., three). Then, for a given hyperparameter configuration, we train our model on *k*–1 folds and evaluate on the remaining fold, repeating this process *k* times. Figure 10-13 illustrates this approach.

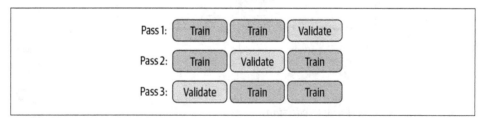

Figure 10-13. k-fold cross-validation

As this figure shows, if we split our data into three folds, our model is first trained on the first and second folds (or splits) of the data, and evaluated on the third fold. We then build the same model with the same hyperparameters on the first and third folds of the data, and evaluate its performance on the second fold. Lastly, we build the model on the second and third folds and evaluate it on the first fold. We then average the performance of those three (or *k*) validation data sets as a proxy of how well this model will perform on unseen data, as every data point had the chance to be part of the validation data set exactly once. Next, we repeat this process for all of our different hyperparameter configurations to identify the optimal one.

Determining the search space of your hyperparameters can be difficult, and often doing a random search of hyperparameters outperforms a structured grid search (*https://oreil.ly/gI7G-*). There are specialized libraries, such as Hyperopt (*http://hyperopt.github.io/hyperopt*), to help you identify the optimal hyperparameter configurations (*https://oreil.ly/7IoxC*), which we touch upon in Chapter 11.

To perform a hyperparameter search in Spark, take the following steps :

1. Define the `estimator` you want to evaluate.

2. Specify which hyperparameters you want to vary, as well as their respective values, using the `ParamGridBuilder` (*https://oreil.ly/qOHrU*).

3. Define an `evaluator` to specify which metric to use to compare the various models.

4. Use the `CrossValidator` (*https://oreil.ly/ygbF8*) to perform cross-validation, evaluating each of the various models.

Let's start by defining our pipeline estimator:

```
# In Python
pipeline = Pipeline(stages = [stringIndexer, vecAssembler, rf])

// In Scala
val pipeline = new Pipeline()
  .setStages(Array(stringIndexer, vecAssembler, rf))
```

For our `ParamGridBuilder`, we'll vary our `maxDepth` to be 2, 4, or 6 and `numTrees` (the number of trees in our random forest) to be 10 or 100. This will give us a grid of 6 (3 x 2) different hyperparameter configurations in total:

```
(maxDepth=2, numTrees=10)
(maxDepth=2, numTrees=100)
(maxDepth=4, numTrees=10)
(maxDepth=4, numTrees=100)
(maxDepth=6, numTrees=10)
(maxDepth=6, numTrees=100)

# In Python
from pyspark.ml.tuning import ParamGridBuilder
paramGrid = (ParamGridBuilder()
            .addGrid(rf.maxDepth, [2, 4, 6])
```

```
    .addGrid(rf.numTrees, [10, 100])
    .build())

// In Scala
import org.apache.spark.ml.tuning.ParamGridBuilder
val paramGrid = new ParamGridBuilder()
  .addGrid(rf.maxDepth, Array(2, 4, 6))
  .addGrid(rf.numTrees, Array(10, 100))
  .build()
```

Now that we have set up our hyperparameter grid, we need to define how to evaluate each of the models to determine which one performed best. For this task we will use the RegressionEvaluator, and we'll use RMSE as our metric of interest:

```
# In Python
evaluator = RegressionEvaluator(labelCol="price",
                                predictionCol="prediction",
                                metricName="rmse")

// In Scala
val evaluator = new RegressionEvaluator()
  .setLabelCol("price")
  .setPredictionCol("prediction")
  .setMetricName("rmse")
```

We will perform our *k*-fold cross-validation using the CrossValidator, which accepts an estimator, evaluator, and estimatorParamMaps so that it knows which model to use, how to evaluate the model, and which hyperparameters to set for the model. We can also set the number of folds we want to split our data into (numFolds=3), as well as setting a seed so we have reproducible splits across the folds (seed=42). Let's then fit this cross-validator to our training data set:

```
# In Python
from pyspark.ml.tuning import CrossValidator

cv = CrossValidator(estimator=pipeline,
                    evaluator=evaluator,
                    estimatorParamMaps=paramGrid,
                    numFolds=3,
                    seed=42)
cvModel = cv.fit(trainDF)

// In Scala
import org.apache.spark.ml.tuning.CrossValidator

val cv = new CrossValidator()
 .setEstimator(pipeline)
 .setEvaluator(evaluator)
 .setEstimatorParamMaps(paramGrid)
 .setNumFolds(3)
 .setSeed(42)
val cvModel = cv.fit(trainDF)
```

The output tells us how long the operation took:

```
Command took 1.07 minutes
```

So, how many models did we just train? If you answered 18 (6 hyperparameter configurations x 3-fold cross-validation), you're close. Once you've identified the optimal hyperparameter configuration, how do you combine those three (or k) models together? While some models might be easy enough to average together, some are not. Therefore, Spark retrains your model on the entire training data set once it has identified the optimal hyperparameter configuration, so in the end we trained 19 models. If you want to retain the intermediate models trained, you can set collectSubModels=True in the CrossValidator.

To inspect the results of the cross-validator, you can take a look at the avgMetrics:

```python
# In Python
list(zip(cvModel.getEstimatorParamMaps(), cvModel.avgMetrics))
```

```scala
// In Scala
cvModel.getEstimatorParamMaps.zip(cvModel.avgMetrics)
```

Here's the output:

```
res1: Array[(org.apache.spark.ml.param.ParamMap, Double)] =
Array(({
    rfr_a132fb1ab6c8-maxDepth: 2,
    rfr_a132fb1ab6c8-numTrees: 10
},303.99522869739343), ({
    rfr_a132fb1ab6c8-maxDepth: 2,
    rfr_a132fb1ab6c8-numTrees: 100
},299.56501993529474), ({
    rfr_a132fb1ab6c8-maxDepth: 4,
    rfr_a132fb1ab6c8-numTrees: 10
},310.63687030886894), ({
    rfr_a132fb1ab6c8-maxDepth: 4,
    rfr_a132fb1ab6c8-numTrees: 100
},294.7369599168999), ({
    rfr_a132fb1ab6c8-maxDepth: 6,
    rfr_a132fb1ab6c8-numTrees: 10
},312.6678169109293), ({
    rfr_a132fb1ab6c8-maxDepth: 6,
    rfr_a132fb1ab6c8-numTrees: 100
},292.101039874209))
```

We can see that the best model from our CrossValidator (the one with the lowest RMSE) had maxDepth=6 and numTrees=100. However, this took a long time to run. In the next section, we will look at how we can decrease the time to train our model while maintaining the same model performance.

Optimizing Pipelines

If your code takes long enough for you to think about improving it, then you should optimize it. In the preceding code, even though each of the models in the cross-validator is technically independent, `spark.ml` actually trains the collection of models sequentially rather than in parallel. In Spark 2.3, a `parallelism` (*https://oreil.ly/7-zyU*) parameter was introduced to solve this problem. This parameter determines the number of models to train in parallel, which themselves are fit in parallel. From the Spark Tuning Guide (*https://oreil.ly/FCXV7*):

> The value of `parallelism` should be chosen carefully to maximize parallelism without exceeding cluster resources, and larger values may not always lead to improved performance. Generally speaking, a value up to 10 should be sufficient for most clusters.

Let's set this value to 4 and see if we can train any faster:

```python
# In Python
cvModel = cv.setParallelism(4).fit(trainDF)
```

```scala
// In Scala
val cvModel = cv.setParallelism(4).fit(trainDF)
```

The answer is yes:

```
Command took 31.45 seconds
```

We've cut the training time in half (from 1.07 minutes to 31.45 seconds), but we can still improve it further! There's another trick we can use to speed up model training: putting the cross-validator inside the pipeline (e.g., `Pipeline(stages=[..., cv]`) instead of putting the pipeline inside the cross-validator (e.g., `CrossValidator(estimator=pipeline, ...)`). Every time the cross-validator evaluates the pipeline, it runs through every step of the pipeline for each model, even if some of the steps don't change, such as the `StringIndexer`. By reevaluating every step in the pipeline, we are learning the same `StringIndexer` mapping over and over again, even though it's not changing.

If instead we put our cross-validator inside our pipeline, then we won't be reevaluating the `StringIndexer` (or any other estimator) each time we try a different model:

```python
# In Python
cv = CrossValidator(estimator=rf,
                    evaluator=evaluator,
                    estimatorParamMaps=paramGrid,
                    numFolds=3,
                    parallelism=4,
                    seed=42)

pipeline = Pipeline(stages=[stringIndexer, vecAssembler, cv])
pipelineModel = pipeline.fit(trainDF)
```

```scala
// In Scala
val cv = new CrossValidator()
  .setEstimator(rf)
  .setEvaluator(evaluator)
  .setEstimatorParamMaps(paramGrid)
  .setNumFolds(3)
  .setParallelism(4)
  .setSeed(42)

val pipeline = new Pipeline()
                  .setStages(Array(stringIndexer, vecAssembler, cv))
val pipelineModel = pipeline.fit(trainDF)
```

This trims five seconds off our training time:

```
Command took 26.21 seconds
```

Thanks to the `parallelism` parameter and rearranging the ordering of our pipeline, that last run was the fastest—and if you apply it to the test data set you'll see that you get the same results. Although these gains were on the order of seconds, the same techniques apply to much larger data sets and models, with correspondingly larger time savings. You can try running this code yourself by accessing the notebook in the book's GitHub repo (*https://github.com/databricks/LearningSparkV2*).

Summary

In this chapter we covered how to build pipelines using Spark MLlib—in particular, its DataFrame-based API package, `spark.ml`. We discussed the differences between transformers and estimators, how to compose them using the Pipeline API, and some different metrics for evaluating models. We then explored how to use cross-validation to perform hyperparameter tuning to deliver the best model, as well as tips for optimizing cross-validation and model training in Spark.

All this sets the context for the next chapter, in which we will discuss deployment strategies and ways to manage and scale machine learning pipelines with Spark.

Managing, Deploying, and Scaling Machine Learning Pipelines with Apache Spark

In the previous chapter, we covered how to build machine learning pipelines with MLlib. This chapter will focus on how to manage and deploy the models you train. By the end of this chapter, you will be able to use MLflow to track, reproduce, and deploy your MLlib models, discuss the difficulties of and trade-offs among various model deployment scenarios, and architect scalable machine learning solutions. But before we discuss deploying models, let's first discuss some best practices for model management to get your models ready for deployment.

Model Management

Before you deploy your machine learning model, you should ensure that you can reproduce and track the model's performance. For us, end-to-end reproducibility of machine learning solutions means that we need to be able to reproduce the code that generated a model, the environment used in training, the data it was trained on, and the model itself. Every data scientist loves to remind you to set your seeds so you can reproduce your experiments (e.g., for the train/test split, when using models with inherent randomness such as random forests). However, there are many more aspects that contribute to reproducibility than just setting seeds, and some of them are much more subtle. Here are a few examples:

Library versioning

When a data scientist hands you their code, they may or may not mention the dependent libraries. While you are able to figure out which libraries are required by going through the error messages, you won't be certain which library versions they used, so you'll likely install the latest ones. But if their code was built on a previous version of a library, which may be taking advantage of some default behavior that differs from the version you installed, using the latest version can cause the code to break or the results to differ (for example, consider how XGBoost (*https://xgboost.readthedocs.io/en/latest*) changed how it handles missing values (*https://oreil.ly/frAKS*) in v0.90).

Data evolution

Suppose you build a model on June 1, 2020, and keep track of all your hyperparameters, libraries, etc. You then try to reproduce the same model on July 1, 2020—but the pipeline breaks or the results differ because the underlying data has changed, which could happen if someone added an extra column or an order of magnitude more data after the initial build.

Order of execution

If a data scientist hands you their code, you should be able to run it top-to-bottom without error. However, data scientists are notorious for running things out of order, or running the same stateful cell multiple times, making their results very difficult to reproduce. (They might also check in a copy of the code with different hyperparameters than those used to train the final model!)

Parallel operations

To maximize throughput, GPUs will run many operations in parallel. However, the order of execution is not always guaranteed, which can lead to nondeterministic outputs. This is a known problem with functions like `tf.reduce_sum()` (*https://oreil.ly/FxNt2*) and when aggregating floating-point numbers (which have limited precision): the order in which you add them may generate slightly different results, which can be exacerbated across many iterations.

An inability to reproduce your experiments can often be a blocker in getting business units to adopt your model or put it into production. While you could build your own in-house tools for tracking your models, data, dependency versions, etc., they may become obsolete, brittle, and take significant development effort to maintain. Equally important is having industry-wide standards for managing models so that they can be easily shared with partners. There are both open source and proprietary tools that can help us with reproducing our machine learning experiments by abstracting away many of these common difficulties. This section will focus on MLflow, as it has the tightest integration with MLlib of the currently available open source model management tools.

MLflow

MLflow (*https://mlflow.org*) is an open source platform that helps developers reproduce and share experiments, manage models, and much more. It provides interfaces in Python, R, and Java/Scala, as well as a REST API. As shown in Figure 11-1, MLflow has four main components:

Tracking

Provides APIs to record parameters, metrics, code versions, models, and artifacts such as plots, and text.

Projects

A standardized format to package your data science projects and their dependencies to run on other platforms. It helps you manage the model training process.

Models

A standardized format to package models to deploy to diverse execution environments. It provides a consistent API for loading and applying models, regardless of the algorithm or library used to build the model.

Registry

A repository to keep track of model lineage, model versions, stage transitions, and annotations.

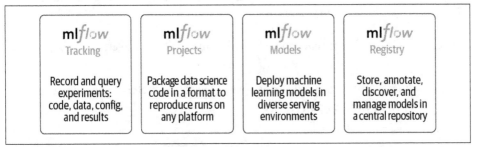

Figure 11-1. MLflow components

Let's track the MLlib model experiments we ran in Chapter 10 for reproducibility. We will then see how the other components of MLflow come into play when we discuss model deployment. To get started with MLflow, simply run `pip install mlflow` on your local host.

Tracking

MLflow Tracking is a logging API that is agnostic to the libraries and environments that actually do the training. It is organized around the concept of *runs*, which are executions of data science code. Runs are aggregated into *experiments*, such that many runs can be part of a given experiment.

The MLflow tracking server can host many experiments. You can log to the tracking server using a notebook, local app, or cloud job, as shown in Figure 11-2.

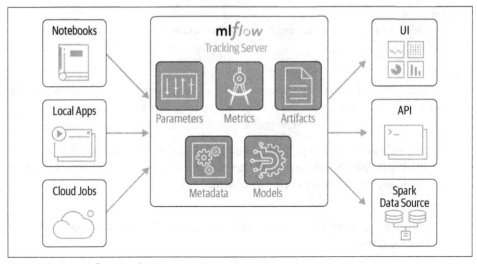

Figure 11-2. MLflow tracking server

Let's examine a few things that can be logged to the tracking server:

Parameters
    Key/value inputs to your code—e.g., hyperparameters like `num_trees` or `max_depth` in your random forest

Metrics
    Numeric values (can update over time)—e.g., RMSE or accuracy values

Artifacts
    Files, data, and models—e.g., `matplotlib` images, or Parquet files

Metadata
    Information about the run, such as the source code that executed the run or the version of the code (e.g., the Git commit hash string for the code version)

Models
    The model(s) you trained

By default, the tracking server records everything to the filesystem, but you can specify a database (*https://oreil.ly/awTsZ*) for faster querying, such as for the parameters and metrics. Let's add MLflow tracking to our random forest code from Chapter 10:

```
# In Python
from pyspark.ml import Pipeline
from pyspark.ml.feature import StringIndexer, VectorAssembler
from pyspark.ml.regression import RandomForestRegressor
from pyspark.ml.evaluation import RegressionEvaluator

filePath = """/databricks-datasets/learning-spark-v2/sf-airbnb/
sf-airbnb-clean.parquet"""
airbnbDF = spark.read.parquet(filePath)
(trainDF, testDF) = airbnbDF.randomSplit([.8, .2], seed=42)

categoricalCols = [field for (field, dataType) in trainDF.dtypes
                   if dataType == "string"]
indexOutputCols = [x + "Index" for x in categoricalCols]
stringIndexer = StringIndexer(inputCols=categoricalCols,
                              outputCols=indexOutputCols,
                              handleInvalid="skip")

numericCols = [field for (field, dataType) in trainDF.dtypes
               if ((dataType == "double") & (field != "price"))]
assemblerInputs = indexOutputCols + numericCols
vecAssembler = VectorAssembler(inputCols=assemblerInputs,
                               outputCol="features")

rf = RandomForestRegressor(labelCol="price", maxBins=40, maxDepth=5,
                           numTrees=100, seed=42)

pipeline = Pipeline(stages=[stringIndexer, vecAssembler, rf])
```

To start logging with MLflow, you will need to start a run using `mlflow.start_run()`. Instead of explicitly calling `mlflow.end_run()`, the examples in this chapter will use a `with` clause to automatically end the run at the end of the `with` block:

```
# In Python
import mlflow
import mlflow.spark
import pandas as pd

with mlflow.start_run(run_name="random-forest") as run:
    # Log params: num_trees and max_depth
    mlflow.log_param("num_trees", rf.getNumTrees())
    mlflow.log_param("max_depth", rf.getMaxDepth())

    # Log model
    pipelineModel = pipeline.fit(trainDF)
    mlflow.spark.log_model(pipelineModel, "model")

    # Log metrics: RMSE and R2
    predDF = pipelineModel.transform(testDF)
    regressionEvaluator = RegressionEvaluator(predictionCol="prediction",
                                              labelCol="price")
    rmse = regressionEvaluator.setMetricName("rmse").evaluate(predDF)
```

```
r2 = regressionEvaluator.setMetricName("r2").evaluate(predDF)
mlflow.log_metrics({"rmse": rmse, "r2": r2})

# Log artifact: feature importance scores
rfModel = pipelineModel.stages[-1]
pandasDF = (pd.DataFrame(list(zip(vecAssembler.getInputCols(),
                                  rfModel.featureImportances)),
                         columns=["feature", "importance"])
            .sort_values(by="importance", ascending=False))

# First write to local filesystem, then tell MLflow where to find that file
pandasDF.to_csv("feature-importance.csv", index=False)
mlflow.log_artifact("feature-importance.csv")
```

Let's examine the MLflow UI, which you can access by running mlflow ui in your terminal and navigating to *http://localhost:5000/*. Figure 11-3 shows a screenshot of the UI.

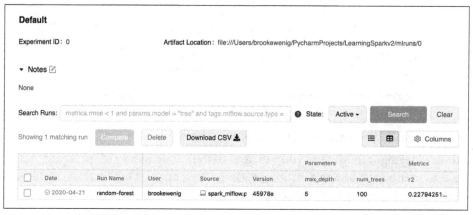

Figure 11-3. The MLflow UI

The UI stores all the runs for a given experiment. You can search across all the runs, filter for those that meet particular criteria, compare runs side by side, etc. If you wish, you can also export the contents as a CSV file to analyze locally. Click on the run in the UI named "random-forest". You should see a screen like Figure 11-4.

Figure 11-4. Random forest run

You'll notice that it keeps track of the source code used for this MLflow run, as well as storing all the corresponding parameters, metrics, etc. You can add notes about this run in free text, as well as tags. You cannot modify the parameters or metrics after the run has finished.

You can also query the tracking server using the `MlflowClient` or REST API:

```python
# In Python
from mlflow.tracking import MlflowClient

client = MlflowClient()
runs = client.search_runs(run.info.experiment_id,
                          order_by=["attributes.start_time desc"],
                          max_results=1)

run_id = runs[0].info.run_id
runs[0].data.metrics
```

This produces the following output:

```
{'r2': 0.22794251914574226, 'rmse': 211.5096898777315}
```

We have hosted this code as an MLflow project (*https://oreil.ly/PleOQ*) in the GitHub repo (*https://github.com/databricks/LearningSparkV2*) for this book, so you can experiment running it with different hyperparameter values for `max_depth` and `num_trees`. The YAML file inside the MLflow project specifies the library dependencies so this code can be run in other environments:

```
# In Python
mlflow.run(
  "https://github.com/databricks/LearningSparkV2/#mlflow-project-example",
  parameters={"max_depth": 5, "num_trees": 100})

# Or on the command line
mlflow run https://github.com/databricks/LearningSparkV2/#mlflow-project-example
-P max_depth=5 -P num_trees=100
```

Now that you have tracked and reproduced your experiments, let's discuss the various deployment options available for your MLlib models.

Model Deployment Options with MLlib

Deploying machine learning models means something different for every organization and use case. Business constraints will impose different requirements for latency, throughput, cost, etc., which dictate which mode of model deployment is suitable for the task at hand—be it batch, streaming, real-time, or mobile/embedded. Deploying models on mobile/embedded systems is outside the scope of this book, so we will focus primarily on the other options. Table 11-1 shows the throughput (*https://oreil.ly/qp1nZ*) and latency (*https://oreil.ly/R7fzj*) trade-offs for these three deployment options for generating predictions. We care about both the number of concurrent requests and the size of those requests, and the resulting solutions will look quite different.

Table 11-1. Batch, streaming, and real-time comparison

	Throughput	Latency	Example application
Batch	High	High (hours to days)	Customer churn prediction
Streaming	Medium	Medium (seconds to minutes)	Dynamic pricing
Real-time	Low	Low (milliseconds)	Online ad bidding

Batch processing generates predictions on a regular schedule and writes the results out to persistent storage to be served elsewhere. It is typically the cheapest and easiest deployment option as you only need to pay for compute during your scheduled run. Batch processing is much more efficient per data point because you accumulate less overhead when amortized across all predictions made. This is particularly the case

---

with Spark, because of the overhead of communicating back and forth between the driver and the executors—you wouldn't want to make predictions one data point at a time! However, its main drawback is latency, as it is typically scheduled with a period of hours or days to generate the next batch of predictions.

Streaming provides a nice trade-off between throughput and latency. You will continuously make predictions on micro-batches of data and get your predictions in seconds to minutes. If you are using Structured Streaming, almost all of your code will look identical to the batch use case, making it easy to go back and forth between these two options. With streaming, you will have to pay for the VMs or computing resources you use to continually stay up and running, and ensure that you have configured the stream properly to be fault tolerant and provide buffering if there are spikes in the incoming data.

Real-time deployment prioritizes latency over throughput and generates predictions in a few milliseconds. Your infrastructure will need to support load balancing and be able to scale to many concurrent requests if there is a large spike in demand (e.g., for online retailers around the holidays). Sometimes when people say "real-time deployment" they mean extracting precomputed predictions in real time, but here we're referring to *generating* model predictions in real time. Real-time deployment is the only option that Spark cannot meet the latency requirements for, so to use it you will need to export your model outside of Spark. For example, if you intend to use a REST endpoint for real-time model inference (say, computing predictions in under 50 ms), MLlib does not meet the latency requirements necessary for this application, as shown in Figure 11-5. You will need to get your feature preparation and model out of Spark, which can be time-consuming and difficult.

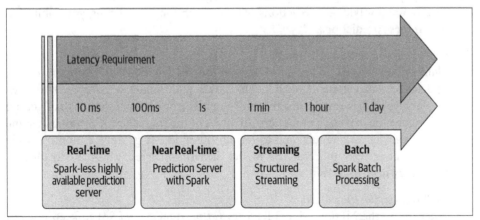

Figure 11-5. Deployment options for MLlib

Before you begin the modeling process, you need to define your model deployment requirements. MLlib and Spark are just a few tools in your toolbox, and you need to understand when and where they should be applied. The remainder of this section discusses the deployment options for MLlib in more depth, and then we'll consider the deployment options with Spark for non-MLlib models.

Batch

Batch deployments represent the majority of use cases for deploying machine learning models, and this is arguably the easiest option to implement. You will run a regular job to generate predictions, and save the results to a table, database, data lake, etc. for downstream consumption. In fact, you have already seen how to generate batch predictions in Chapter 10 with MLlib. MLlib's model.transform() will apply the model in parallel to all partitions of your DataFrame:

```
# In Python
# Load saved model with MLflow
import mlflow.spark
pipelineModel = mlflow.spark.load_model(f"runs:/{run_id}/model")

# Generate predictions
inputDF = spark.read.parquet("/databricks-datasets/learning-spark-v2/
  sf-airbnb/sf-airbnb-clean.parquet")

predDF = pipelineModel.transform(inputDF)
```

A few things to keep in mind with batch deployments are:

How frequently will you generate predictions?
> There is a trade-off between latency and throughput. You will get higher throughput batching many predictions together, but then the time it takes to receive any individual predictions will be much longer, delaying your ability to act on these predictions.

How often will you retrain the model?
> Unlike libraries like sklearn or TensorFlow, MLlib does not support online updates or warm starts. If you'd like to retrain your model to incorporate the latest data, you'll have to retrain the entire model from scratch, rather than getting to leverage the existing parameters. In terms of the frequency of retraining, some people will set up a regular job to retrain the model (e.g., once a month), while others will actively monitor the model drift (*https://oreil.ly/aX4dT*) to identify when they need to retrain.

How will you version the model?
> You can use the MLflow Model Registry (*https://oreil.ly/D5LR6*) to keep track of the models you are using and control how they are transitioned to/from staging, production, and archived. You can see a screenshot of the Model Registry in Figure 11-6. You can use the Model Registry with the other deployment options too.

Figure 11-6. MLflow Model Registry

In addition to using the MLflow UI to manage your models, you can also manage them programmatically. For example, once you have registered your production model, it has a consistent URI that you can use to retrieve the latest version:

```
# Retrieve latest production model
model_production_uri = f"models:/{model_name}/production"
model_production = mlflow.spark.load_model(model_production_uri)
```

Streaming

Instead of waiting for an hourly or nightly job to process your data and generate predictions, Structured Streaming can continuously perform inference on incoming data. While this approach is more costly than a batch solution as you have to continually pay for compute time (and get lower throughput), you get the added benefit of generating predictions more frequently so you can act on them sooner. Streaming solutions in general are more complicated to maintain and monitor than batch solutions, but they offer lower latency.

With Spark it's very easy to convert your batch predictions to streaming predictions, and practically all of the code is the same. The only difference is that when you read in the data, you need to use `spark.readStream()` rather than `spark.read()` and change the source of the data. In the following example we are going to simulate reading in streaming data by streaming in a directory of Parquet files. You'll notice that we are specifying a `schema` even though we are working with Parquet files. This is because we need to define the schema a priori when working with streaming data. In this example, we will use the random forest model trained on our Airbnb data set from the previous chapter to perform these streaming predictions. We will load in the saved model using MLflow. We have partitioned the source file into one hundred small Parquet files so you can see the output changing at every trigger interval:

```
# In Python
# Load saved model with MLflow
pipelineModel = mlflow.spark.load_model(f"runs:/{run_id}/model")
```

```
# Set up simulated streaming data
repartitionedPath = "/databricks-datasets/learning-spark-v2/sf-airbnb/
  sf-airbnb-clean-100p.parquet"
schema = spark.read.parquet(repartitionedPath).schema

streamingData = (spark
                .readStream
                .schema(schema) # Can set the schema this way
                .option("maxFilesPerTrigger", 1)
                .parquet(repartitionedPath))

# Generate predictions
streamPred = pipelineModel.transform(streamingData)
```

After you generate these predictions, you can write them out to any target location for retrieval later (refer to Chapter 8 for Structured Streaming tips). As you can see, the code is virtually unchanged between the batch and streaming scenarios, making MLlib a great solution for both. However, depending on the latency demands of your task, MLlib may not be the best choice. With Spark there is significant overhead involved in generating the query plan and communicating the task and results between the driver and the worker. Thus, if you need really low-latency predictions, you'll need to export your model out of Spark.

Near Real-Time

If your use case requires predictions on the order of hundreds of milliseconds to seconds, you could build a prediction server that uses MLlib to generate the predictions. While this is not an ideal use case for Spark because you are processing very small amounts of data, you'll get lower latency than with streaming or batch solutions.

Model Export Patterns for Real-Time Inference

There are some domains where real-time inference is required, including fraud detection, ad recommendation, and the like. While making predictions with a small number of records may achieve the low latency required for real-time inference, you will need to contend with load balancing (handling many concurrent requests) as well as geolocation in latency-critical tasks. There are popular managed solutions, such as AWS SageMaker (*https://aws.amazon.com/sagemaker*) and Azure ML (*https://oreil.ly/OzEnY*), that provide low-latency model serving solutions. In this section we'll show you how to export your MLlib models so they can be deployed to those services.

One way to export your model out of Spark is to reimplement the model natively in Python, C, etc. While it may seem simple to extract the coefficients of the model, exporting all the feature engineering and preprocessing steps along with them (OneHo tEncoder, VectorAssembler, etc.) quickly gets troublesome and is very error-prone.

There are a few open source libraries, such as MLeap and ONNX (*https://onnx.ai*), that can help you automatically export a supported subset of the MLlib models to remove their dependency on Spark. However, as of the time of this writing the company that developed MLeap is no longer supporting it. Nor does MLeap yet support Scala 2.12/Spark 3.0.

ONNX (Open Neural Network Exchange), on the other hand, has become the de facto open standard for machine learning interoperability. Some of you might recall other ML interoperability formats, like PMML (Predictive Model Markup Language), but those never gained quite the same traction as ONNX has now. ONNX is very popular in the deep learning community as a tool that allows developers to easily switch between libraries and languages, and at the time of this writing it has experimental support for MLlib.

Instead of exporting MLlib models, there are other third-party libraries that integrate with Spark that are convenient to deploy in real-time scenarios, such as XGBoost (*https://oreil.ly/7-iZJ*) and H2O.ai's Sparkling Water (*https://oreil.ly/yhKP9*) (whose name is derived from a combination of H2O and Spark).

XGBoost is one of the most successful algorithms in Kaggle competitions (*https://www.kaggle.com*) for structured data problems, and it's a very popular library among data scientists. Although XGBoost is not technically part of MLlib, the XGBoost4J-Spark library (*https://oreil.ly/XGg5c*) allows you to integrate distributed XGBoost into your MLlib pipelines. A benefit of XGBoost is the ease of deployment: after you train your MLlib pipeline, you can extract the XGBoost model and save it as a non-Spark model for serving in Python, as demonstrated here:

```scala
// In Scala
val xgboostModel =
  xgboostPipelineModel.stages.last.asInstanceOf[XGBoostRegressionModel]
xgboostModel.nativeBooster.saveModel(nativeModelPath)
```

```python
# In Python
import xgboost as xgb
bst = xgb.Booster({'nthread': 4})
bst.load_model("xgboost_native_model")
```

 At the time of this writing, the distributed XGBoost API is only available in Java/Scala. A full example is included in the book's GitHub repo (*https://github.com/databricks/LearningSparkV2*).

Now that you have learned about the different ways of exporting MLlib models for use in real-time serving environments, let's discuss how we can leverage Spark for non-MLlib models.

Leveraging Spark for Non-MLlib Models

As mentioned previously, MLlib isn't always the best solution for your machine learning needs. It may not meet super low-latency inference requirements or have built-in support for the algorithm you'd like to use. For these cases, you can still leverage Spark, but not MLlib. In this section, we will discuss how you can use Spark to perform distributed inference of single-node models using Pandas UDFs, perform hyperparameter tuning, and scale feature engineering.

Pandas UDFs

While MLlib is fantastic for distributed training of models, you are not limited to just using MLlib for making batch or streaming predictions with Spark—you can create custom functions to apply your pretrained models at scale, known as user-defined functions (UDFs, covered in Chapter 5). A common use case is to build a scikit-learn or TensorFlow model on a single machine, perhaps on a subset of your data, but perform distributed inference on the entire data set using Spark.

If you define your own UDF to apply a model to each record of your DataFrame in Python, opt for pandas UDFs (*https://oreil.ly/ww2_S*) for optimized serialization and deserialization, as discussed in Chapter 5. However, if your model is very large, then there is high overhead for the Pandas UDF to repeatedly load the same model for every batch in the same Python worker process. In Spark 3.0, Pandas UDFs can accept an iterator of `pandas.Series` or `pandas.DataFrame` so that you can load the model only once instead of loading it for every series in the iterator. For more details on what's new in Apache Spark 3.0 with Pandas UDFs, see Chapter 12.

> If the workers cached the model weights after loading it for the first time, subsequent calls of the same UDF with the same model loading will become significantly faster.

In the following example, we will use `mapInPandas()`, introduced in Spark 3.0, to apply a `scikit-learn` model to our Airbnb data set. `mapInPandas()` takes an iterator of `pandas.DataFrame` as input, and outputs another iterator of `pandas.DataFrame`. It's flexible and easy to use if your model requires all of your columns as input, but it requires serialization/deserialization of the whole DataFrame (as it is passed to its input). You can control the size of each `pandas.DataFrame` with the `spark.sql.execution.arrow.maxRecordsPerBatch` config. A full copy of the code to generate the model is available in this book's GitHub repo (*https://github.com/databricks/Learning SparkV2*), but here we will just focus on loading the saved `scikit-learn` model from MLflow and applying it to our Spark DataFrame:

```python
# In Python
import mlflow.sklearn
import pandas as pd

def predict(iterator):
    model_path = f"runs:/{run_id}/random-forest-model"
    model = mlflow.sklearn.load_model(model_path) # Load model
    for features in iterator:
        yield pd.DataFrame(model.predict(features))

df.mapInPandas(predict, "prediction double").show(3)

+-----------------+
|       prediction|
+-----------------+
| 90.4355866254844|
|255.3459534312323|
| 499.625544914651|
+-----------------+
```

In addition to applying models at scale using a Pandas UDF, you can also use them to parallelize the process of building many models. For example, you might want to build a model for each IoT device type to predict time to failure. You can use `pyspark.sql.GroupedData.applyInPandas()` (introduced in Spark 3.0) for this task. The function takes a `pandas.DataFrame` and returns another `pandas.DataFrame`. The book's GitHub repo contains a full example of the code to build a model per IoT device type and track the individual models with MLflow; just a snippet is included here for brevity:

```python
# In Python
df.groupBy("device_id").applyInPandas(build_model, schema=trainReturnSchema)
```

The `groupBy()` will cause a full shuffle of your data set, and you need to ensure that your model and the data for each group can fit on a single machine. Some of you might be familiar with `pyspark.sql.GroupedData.apply()` (e.g., `df.groupBy("device_id").apply(build_model)`), but that API will be deprecated in future releases of Spark in favor of `pyspark.sql.GroupedData.applyInPandas()`.

Now that you have seen how to apply UDFs to perform distributed inference and parallelize model building, let's look at how to use Spark for distributed hyperparameter tuning.

Spark for Distributed Hyperparameter Tuning

Even if you do not intend to do distributed inference or do not need MLlib's distributed training capabilities, you can still leverage Spark for distributed hyperparameter tuning. This section will cover two open source libraries in particular: Joblib and Hyperopt.

Joblib

According to its documentation, Joblib (*https://github.com/joblib/joblib*) is "a set of tools to provide lightweight pipelining in Python." It has a Spark backend to distribute tasks on a Spark cluster. Joblib can be used for hyperparameter tuning as it automatically broadcasts a copy of your data to all of your workers, which then create their own models with different hyperparameters on their copies of the data. This allows you to train and evaluate multiple models in parallel. You still have the fundamental limitation that a single model and all the data have to fit on a single machine, but you can trivially parallelize the hyperparameter search, as shown in Figure 11-7.

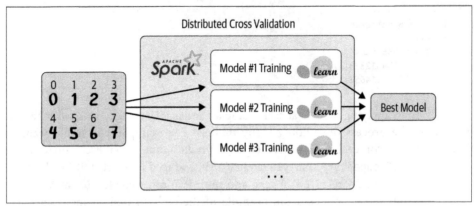

Figure 11-7. Distributed hyperparameter search

To use Joblib, install it via `pip install joblibspark`. Ensure you are using `scikit-learn` version 0.21 or later and `pyspark` version 2.4.4 or later. An example of how to do distributed cross-validation is shown here, and the same approach will work for distributed hyperparameter tuning as well:

```python
# In Python
from sklearn.utils import parallel_backend
from sklearn.ensemble import RandomForestRegressor
from sklearn.model_selection import train_test_split
from sklearn.model_selection import GridSearchCV
import pandas as pd
from joblibspark import register_spark

register_spark() # Register Spark backend

df = pd.read_csv("/dbfs/databricks-datasets/learning-spark-v2/sf-airbnb/
  sf-airbnb-numeric.csv")
X_train, X_test, y_train, y_test = train_test_split(df.drop(["price"], axis=1),
  df[["price"]].values.ravel(), random_state=42)

rf = RandomForestRegressor(random_state=42)
param_grid = {"max_depth": [2, 5, 10], "n_estimators": [20, 50, 100]}
```

```
gscv = GridSearchCV(rf, param_grid, cv=3)

with parallel_backend("spark", n_jobs=3):
    gscv.fit(X_train, y_train)

print(gscv.cv_results_)
```

See the `scikit-learn` GridSearchCV documentation (*https://oreil.ly/zjuSD*) for an explanation of the parameters returned from the cross-validator.

Hyperopt

Hyperopt (*https://oreil.ly/N9TVh*) is a Python library for "serial and parallel optimization over awkward search spaces, which may include real-valued, discrete, and conditional dimensions." You can install it via `pip install hyperopt`. There are two main ways to scale Hyperopt with Apache Spark (*https://oreil.ly/D07fV*):

- Using single-machine Hyperopt with a distributed training algorithm (e.g., MLlib)
- Using distributed Hyperopt with single-machine training algorithms with the `SparkTrials` class

For the former case, there is nothing special you need to configure to use MLlib with Hyperopt versus any other library. So, let's take a look at the latter case: distributed Hyperopt with single-node models. Unfortunately, you can't combine distributed hyperparameter evaluation with distributed training models at the time of this writing. The full code example for parallelizing the hyperparameter search for a Keras (*https://oreil.ly/XbHSG*) model can be found in the book's GitHub repo (*https://github.com/databricks/LearningSparkV2*); just a snippet is included here to illustrate the key components of Hyperopt:

```
# In Python
import hyperopt

best_hyperparameters = hyperopt.fmin(
    fn = training_function,
    space = search_space,
    algo = hyperopt.tpe.suggest,
    max_evals = 64,
    trials = hyperopt.SparkTrials(parallelism=4))
```

`fmin()` generates new hyperparameter configurations to use for your `training_func` `tion` and passes them to `SparkTrials`. `SparkTrials` runs batches of these training tasks in parallel as a single-task Spark job on each Spark executor. When the Spark task is done, it returns the results and the corresponding loss to the driver. Hyperopt uses these new results to compute better hyperparameter configurations for future tasks. This allows for massive scale-out of hyperparameter tuning. MLflow also integrates with Hyperopt, so you can track the results of all the models you've trained as part of your hyperparameter tuning.

An important parameter for `SparkTrials` is `parallelism`. This determines the maximum number of trials to evaluate concurrently. If `parallelism=1`, then you are training each model sequentially, but you might get better models by making full use of adaptive algorithms. If you set `parallelism=max_evals` (the total number of models to train), then you are just doing a random search. Any number between 1 and `max_evals` allows you to have a trade-off between scalability and adaptiveness. By default, `parallelism` is set to the number of Spark executors. You can also specify a `timeout` to limit the maximum number of seconds that `fmin()` is allowed to take.

Even if MLlib isn't suitable for your problem, hopefully you can see the value of using Spark in any of your machine learning tasks.

Koalas

Pandas (*https://oreil.ly/Z9rcQ*) is a very popular data analysis and manipulation library in Python, but it is limited to running on a single machine. Koalas (*https://github.com/databricks/koalas*) is an open source library that implements the Pandas DataFrame API on top of Apache Spark, easing the transition from Pandas to Spark. You can install it with `pip install koalas`, and then simply replace any pd (Pandas) logic in your code with ks (Koalas). This way, you can scale up your analyses with Pandas without needing to entirely rewrite your codebase in PySpark. Here is an example of how to change your Pandas code to Koalas (you'll need to have PySpark already installed):

```
# In pandas
import pandas as pd
pdf = pd.read_csv(csv_path, header=0, sep=";", quotechar='"')
pdf["duration_new"] = pdf["duration"] + 100

# In koalas
import databricks.koalas as ks
kdf = ks.read_csv(file_path, header=0, sep=";", quotechar='"')
kdf["duration_new"] = kdf["duration"] + 100
```

While Koalas aims to implement all Pandas features eventually, not all of them are implemented yet. If there is functionality that you need that Koalas does not provide, you can always switch to using the Spark APIs by calling `kdf.to_spark()`. Alternatively, you can bring the data to the driver by calling `kdf.to_pandas()` and use the Pandas API (be careful the data set isn't too large or you will crash the driver!).

Summary

In this chapter, we covered a variety of best practices for managing and deploying machine learning pipelines. You saw how MLflow can help you track and reproduce experiments and package your code and its dependencies to deploy elsewhere. We also discussed the main deployment options—batch, streaming, and real-time—and their associated trade-offs. MLlib is a fantastic solution for large-scale model training and batch/streaming use cases, but it won't beat a single-node model for real-time inference on small data sets. Your deployment requirements directly impact the types of models and frameworks that you can use, and it is critical to discuss these requirements before you begin your model building process.

In the next chapter, we will highlight a handful of key new features in Spark 3.0 and how you can incorporate them into your Spark workloads.

Epilogue: Apache Spark 3.0

At the time we were writing this book, Apache Spark 3.0 had not yet been officially released; it was still under development, and we got to work with Spark 3.0.0-preview2. All the code samples in this book have been tested against Spark 3.0.0-preview2, and they should work no differently with the official Spark 3.0 release. Whenever possible in the chapters, where relevant, we mentioned when features were new additions or behaviors in Spark 3.0. In this chapter, we survey the changes.

The bug fixes and feature enhancements are numerous, so for brevity, we highlight just a selection of the notable changes and features pertaining to Spark components. Some of the new features are, under the hood, advanced and beyond the scope of this book, but we mention them here so you can explore them when the release is generally available.

Spark Core and Spark SQL

Let's first consider what's new under the covers. A number of changes have been introduced in Spark Core and the Spark SQL engine to help speed up queries. One way to expedite queries is to read less data using dynamic partition pruning. Another is to adapt and optimize query plans during execution.

Dynamic Partition Pruning

The idea behind dynamic partition pruning (DPP) (*https://oreil.ly/fizdc*) is to skip over the data you don't need in a query's results. The typical scenario where DPP is optimal is when you are joining two tables: a fact table (partitioned over multiple columns) and a dimension table (nonpartitioned), as shown in Figure 12-1. Normally, the filter is on the nonpartitioned side of the table (Date, in our case). For example, consider this common query over two tables, Sales and Date:

```
-- In SQL
SELECT * FROM Sales JOIN ON Sales.date = Date.date
```

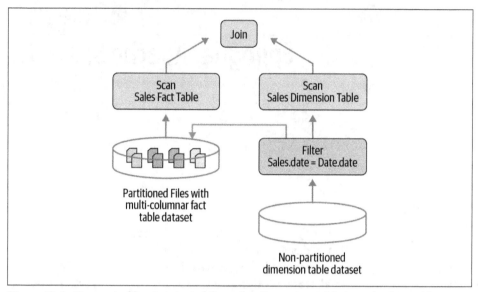

Figure 12-1. Dynamic filter is injected from the dimension table into the fact table

The key optimization technique in DPP is to take the result of the filter from the dimension table and inject it into the fact table as part of the scan operation to limit the data read, as shown in Figure 12-1.

Consider a case where the dimension table is smaller than the fact table and we perform a join, as shown in Figure 12-2. In this case, Spark most likely will do a broadcast join (discussed in Chapter 7). During this join, Spark will conduct the following steps to minimize the amount of data scanned from the larger fact table:

1. On the dimension side of the join, Spark will build a hash table from the dimension table, also known as the build relation, as part of this filter query.

2. Spark will plug the result of this query into the hash table and assign it to a broadcast variable, which is distributed to all executors involved in this join operation.

3. On each executor, Spark will probe the broadcasted hash table to determine what corresponding rows to read from the fact table.

4. Finally, Spark will inject this filter dynamically into the file scan operation of the fact table and reuse the results from the broadcast variable. This way, as part of the file scan operation on the fact table, only the partitions that match the filter are scanned and only the data needed is read.

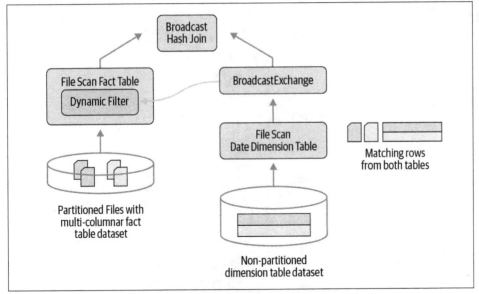

Figure 12-2. Spark injects a dimension table filter into the fact table during a broadcast join

Enabled by default so that you don't have to explicitly configure it, all this happens dynamically when you perform joins between two tables. With the DPP optimization, Spark 3.0 can work much better with star-schema queries.

Adaptive Query Execution

Another way Spark 3.0 optimizes query performance is by adapting its physical execution plan at runtime. *Adaptive Query Execution (AQE)* (*https://oreil.ly/mO8Ua*) reoptimizes and adjusts query plans based on runtime statistics collected in the process of query execution. It attempts to to do the following at runtime:

- Reduce the number of reducers in the shuffle stage by decreasing the number of shuffle partitions.
- Optimize the physical execution plan of the query, for example by converting a SortMergeJoin into a BroadcastHashJoin where appropriate.
- Handle data skew during a join.

All these adaptive measures take place during the execution of the plan at runtime, as shown in Figure 12-3. To use AQE in Spark 3.0, set the configuration spark.sql.adaptive.enabled to true.

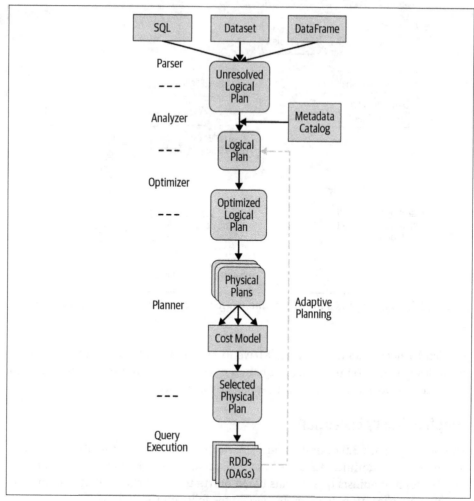

Figure 12-3. AQE reexamines and reoptimizes the execution plan at runtime

The AQE framework

Spark operations in a query are pipelined and executed in parallel processes, but a shuffle or broadcast exchange breaks this pipeline, because the output of one stage is needed as input to the next stage (see "Step 3: Understanding Spark Application Concepts" on page 25 in Chapter 2). These breaking points are called *materialization points* in a query stage, and they present an opportunity to reoptimize and reexamine the query, as illustrated in Figure 12-4.

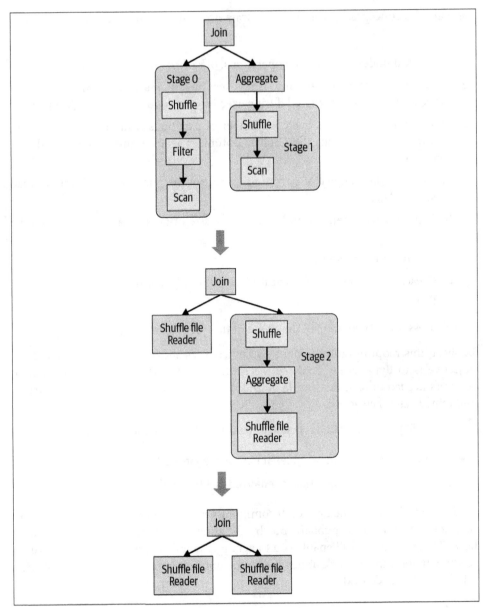

Figure 12-4. A query plan reoptimized in the AQE framework

Here are the conceptual steps the AQE framework iterates over, as depicted in this figure:

1. All the leaf nodes, such as scan operations, of each stage are executed.

2. Once the materialization point finishes executing, it's marked as complete, and all the relevant statistics garnered during execution are updated in its logical plan.

3. Based on these statistics, such as number of partitions read, bytes of data read, etc., the framework runs the Catalyst optimizer again to understand whether it can:

   a. Coalesce the number of partitions to reduce the number of reducers to read shuffle data.

   b. Replace a sort merge join, based on the size of tables read, with a broadcast join.

   c. Try to remedy a skew join.

   d. Create a new optimized logical plan, followed by a new optimized physical plan.

This process is repeated until all the stages of the query plan are executed.

In short, this reoptimization is done dynamically, as shown in Figure 12-3, and the objective is to dynamically coalesce the shuffle partitions, decrease the number of reducers needed to read the shuffle output data, switch join strategies if appropriate, and remedy any skew joins.

Two Spark SQL configurations dictate how AQE will reduce the number of reducers:

- `spark.sql.adaptive.coalescePartitions.enabled` (set to `true`)
- `spark.sql.adaptive.skewJoin.enabled` (set to `true`)

At the time of writing, the Spark 3.0 community blog, documentation, and examples had not been published publicly, but by the time of publication they should have been. These resources will enable you to get more detailed information if you wish to see how these features work under the hood—including on how you can inject SQL join hints, discussed next.

SQL Join Hints

Adding to the existing BROADCAST hints for joins, Spark 3.0 adds join hints for all Spark join strategies (*https://oreil.ly/GqlqH*) (see "A Family of Spark Joins" on page 187 in Chapter 7). Examples are provided here for each type of join.

Shuffle sort merge join (SMJ)

With these new hints, you can suggest to Spark that it perform a `SortMergeJoin` when joining tables `a` and `b` or `customers` and `orders`, as shown in the following examples. You can add one or more hints to a `SELECT` statement inside `/*+ ... */` comment blocks:

```
SELECT /*+ MERGE(a, b) */ id FROM a JOIN b ON a.key = b.key
SELECT /*+ MERGE(customers, orders) */ * FROM customers, orders WHERE
    orders.custId = customers.custId
```

Broadcast hash join (BHJ)

Similarly, for a broadcast hash join, you can provide a hint to Spark that you prefer a broadcast join. For example, here we broadcast table `a` to join with table `b` and table `customers` to join with table `orders`:

```
SELECT /*+ BROADCAST(a) */ id FROM a JOIN b ON a.key = b.key
SELECT /*+ BROADCAST(customers) */ * FROM customers, orders WHERE
    orders.custId = customers.custId
```

Shuffle hash join (SHJ)

You can offer hints in a similar way to perform shuffle hash joins, though this is less commonly encountered than the previous two supported join strategies:

```
SELECT /*+ SHUFFLE_HASH(a, b) */ id FROM a JOIN b ON a.key = b.key
SELECT /*+ SHUFFLE_HASH(customers, orders) */ * FROM customers, orders WHERE
    orders.custId = customers.custId
```

Shuffle-and-replicate nested loop join (SNLJ)

Finally, the shuffle-and-replicate nested loop join adheres to a similar form and syntax:

```
SELECT /*+ SHUFFLE_REPLICATE_NL(a, b) */ id FROM a JOIN b
```

Catalog Plugin API and DataSourceV2

Not to be confined only to the Hive metastore and catalog, Spark 3.0's experimental DataSourceV2 API extends the Spark ecosystem and affords developers three core capabilities. Specifically, it:

- Enables plugging in an external data source for catalog and table management
- Supports predicate pushdown to additional data sources with supported file formats like ORC, Parquet, Kafka, Cassandra, Delta Lake, and Apache Iceberg.
- Provides unified APIs for streaming and batch processing of data sources for sinks and sources

Aimed at developers who want to extend Spark's ability to use external sources and sinks, the Catalog API provides both SQL and programmatic APIs to create, alter, load, and drop tables from the specified pluggable catalog. The catalog provides a hierarchical abstraction of functionalities and operations performed at different levels, as shown in Figure 12-5.

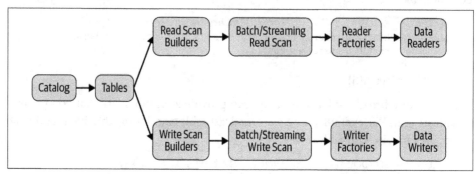

Figure 12-5. Catalog plugin API's hierarchical level of functionality

The initial interaction between Spark and a specific connector is to resolve a relation to its actual `Table` object. `Catalog` defines how to look up tables in this connector. Additionally, `Catalog` can define how to modify its own metadata, thus enabling operations (*https://oreil.ly/TrscV*) like `CREATE TABLE`, `ALTER TABLE`, etc.

For example, in SQL you can now issue commands to create namespaces for your catalog. To use a pluggable catalog, enable the following configs in your *spark-defaults.conf* file:

```
spark.sql.catalog.ndb_catalog com.ndb.ConnectorImpl # connector implementation
spark.sql.catalog.ndb_catalog.option1  value1
spark.sql.catalog.ndb_catalog.option2  value2
```

Here, the connector to the data source catalog has two options: option1->value1 and option2->value2. Once they've been defined, application users in Spark or SQL can use the `DataFrameReader` and `DataFrameWriter` API methods or Spark SQL commands with these defined options as methods for data source manipulation. For example:

```
-- In SQL
SHOW TABLES ndb_catalog;
CREATE TABLE ndb_catalog.table_1;
SELECT * from ndb_catalog.table_1;
ALTER TABLE ndb_catalog.table_1

// In Scala
df.writeTo("ndb_catalog.table_1")
val dfNBD = spark.read.table("ndb_catalog.table_1")
  .option("option1", "value1")
  .option("option2", "value2")
```

While these catalog plugin APIs extend Spark's ability to utilize external data sources as sinks and sources, they are still experimental and should not be used in production. A detailed guide to their use is beyond the scope of this book, but we encourage you to check the release documentation for additional information if you want to write a custom connector to an external data source as a catalog to manage your external tables and their associated metadata.

The preceding code snippets are examples of what your code may look like after you have defined and implemented your catalog connectors and populated them with data.

Accelerator-Aware Scheduler

Project Hydrogen (*https://oreil.ly/Jk4rA*), a community initiative to bring AI and big data together, has three major goals: implementing barrier execution mode, accelerator-aware scheduling, and optimized data exchange. A basic implementation of barrier execution mode (*https://oreil.ly/RDyb1*) was introduced in Apache Spark 2.4.0. In Spark 3.0, a basic scheduler (*https://oreil.ly/9TOyT*) has been implemented to take advantage of hardware accelerators such as GPUs on target platforms where Spark is deployed in standalone mode, YARN, or Kubernetes.

For Spark to take advantage of these GPUs in an organized way for specialized workloads that use them, you have to specify the hardware resources available via configs. Your application can then discover them with the help of a discovery script. Enabling GPU use is a three-step process in your Spark application:

1. Write a discovery script that discovers the addresses of the underlying GPUs available on each Spark executor. This script is set in the following Spark configuration:

   ```
   spark.worker.resource.gpu.discoveryScript=/path/to/script.sh
   ```

2. Set up configuration for your Spark executors to use these discovered GPUs:

   ```
   spark.executor.resource.gpu.amount=2
   spark.task.resource.gpu.amount=1
   ```

3. Write RDD code to leverage these GPUs for your task:

   ```
   import org.apache.spark.BarrierTaskContext
   val rdd = ...
   rdd.barrier.mapPartitions { it =>
     val context = BarrierTaskContext.getcontext.barrier()
     val gpus = context.resources().get("gpu").get.addresses
     // launch external process that leverages GPU
   ```

```
launchProcess(gpus)
    }
```

 These steps are still experimental, and further development will continue in future Spark 3.x releases to support seamless discovery of GPU resources, both at the command line (with spark-submit) and at the Spark task level.

Structured Streaming

To inspect how your Structured Streaming jobs fare with the ebb and flow of data during the course of execution, the Spark 3.0 UI has a new Structured Streaming tab alongside the other tabs we explored in Chapter 7. This tab offers two sets of statistics: aggregate information about completed streaming query jobs (Figure 12-6) and detailed statistics about the streaming queries, including the input rate, process rate, number of input rows, batch duration, and operation duration (Figure 12-7).

Streaming Query

▼Active Streaming Queries (4)

Name	Status	Id	Run ID	Start Time	Duration	Avg Input /sec	Avg Process /sec	Lastest Batch
display_query_1	RUNNING	0b23d2af-7394-4cc7-9dd0-021d830e77fd	456c184b-46fb-4ce4-87e5-5955d0afd563	2020/06/03 23:01:16	3 minutes 5 seconds	33.98	35.72	24
display_query_4	RUNNING	8311ea6a-07a3-40cf-b469-4aa58e71bc35	0afb8817-691f-416b-9366-ff935138ab33	2020/06/03 23:04:08	14 seconds 385 ms	52.85	61.66	7
display_query_2	RUNNING	156e2285-fe84-4840-9383-824e0d9c5250	88c7a058-85fc-4d10-bdb0-4e1d9db5c8f1	2020/06/03 23:03:43	38 seconds 572 ms	56.29	58.60	19
display_query_3	RUNNING	dc7ea927-5631-4325-05da	b9945a08-0c4f-4166-8b7a	2020/06/03 23:03:55	26 seconds 871 ms	52.50	57.03	13

Figure 12-6. Structured Streaming tab showing aggregate statistics of a completed streaming job

The Figure 12-7 screenshot was taken with Spark 3.0.0-preview2; with the final release, you should see the query name and ID in the name identifier on the UI page.

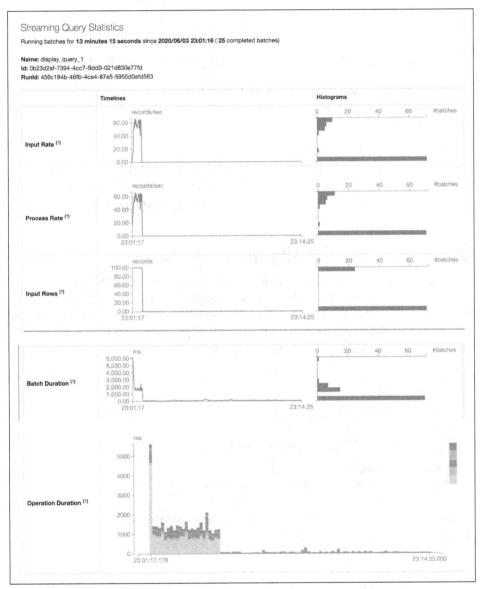

Figure 12-7. Showing detailed statistics of a completed streaming job

No configuration is required (*https://oreil.ly/wP1QB*); all configurations works straight out of the Spark 3.0 installation, with the following defaults:

- `spark.sql.streaming.ui.enabled=true`
- `spark.sql.streaming.ui.retainedProgressUpdates=100`
- `spark.sql.streaming.ui.retainedQueries=100`

PySpark, Pandas UDFs, and Pandas Function APIs

Spark 3.0 requires pandas version 0.23.2 or higher to employ any pandas-related methods, such as `DataFrame.toPandas()` or `SparkSession.createDataFrame(pandas.DataFrame)`.

Furthermore, it requires PyArrow version 0.12.1 or later to use PyArrow functionality such as `pandas_udf()`, `DataFrame.toPandas()`, and `SparkSession.createDataFrame(pandas.DataFrame)` with the `spark.sql.execution.arrow.enabled` configuration set to `true`. The next section will introduce new features in Pandas UDFs.

Redesigned Pandas UDFs with Python Type Hints

The Pandas UDFs in Spark 3.0 were redesigned by leveraging Python type hints (*https://oreil.ly/tAEA9*). This enables you to naturally express UDFs without requiring the evaluation type. Pandas UDFs are now more "Pythonic" and can themselves define what the UDF is supposed to input and output, rather than you specifying it via, for example, `@pandas_udf("long", PandasUDFType.SCALAR)` as you did in Spark 2.4.

Here's an example:

```
# Pandas UDFs in Spark 3.0
import pandas as pd
from pyspark.sql.functions import pandas_udf

@pandas_udf("long")
def pandas_plus_one(v: pd.Series) -> pd.Series:
  return v + 1
```

This new format provides several benefits, such as easier static analysis. You can apply the new UDFs in the same way as before:

```
df = spark.range(3)
df.withColumn("plus_one", pandas_plus_one("id")).show()

+---+--------+
| id|plus_one|
+---+--------+
|  0|       1|
|  1|       2|
|  2|       3|
+---+--------+
```

Iterator Support in Pandas UDFs

Pandas UDFs are very commonly used to load a model and perform distributed inference for single-node machine learning and deep learning models. However, if a model is very large, then there is high overhead for the Pandas UDF to repeatedly load the same model for every batch in the same Python worker process.

In Spark 3.0, Pandas UDFs can accept an iterator (*https://oreil.ly/FboVn*) of pandas.Series or pandas.DataFrame, as shown here:

```
from typing import Iterator

@pandas_udf('long')
def pandas_plus_one(iterator: Iterator[pd.Series]) -> Iterator[pd.Series]:
    return map(lambda s: s + 1, iterator)

df.withColumn("plus_one", pandas_plus_one("id")).show()

+---+--------+
| id|plus_one|
+---+--------+
|  0|       1|
|  1|       2|
|  2|       3|
+---+--------+
```

With this support, you can load the model only once instead of loading it for every series in the iterator. The following pseudocode illustrates how to do this:

```
@pandas_udf(...)
def predict(iterator):
  model = ... # load model
  for features in iterator:
    yield model.predict(features)
```

New Pandas Function APIs

Spark 3.0 introduces a few new types of Pandas UDFs that are useful when you want to apply a function against an entire DataFrame instead of column-wise, such as `mapInPandas()`, introduced in Chapter 11. These take an iterator of `pandas.DataFrame` as input and output another iterator of `pandas.DataFrame`:

```
def pandas_filter(
    iterator: Iterator[pd.DataFrame]) -> Iterator[pd.DataFrame]:
  for pdf in iterator:
    yield pdf[pdf.id == 1]

df.mapInPandas(pandas_filter, schema=df.schema).show()

+---+
| id|
+---+
|  1|
+---+
```

You can control the size of the `pandas.DataFrame` by specifying it in the `spark.sql.execution.arrow.maxRecordsPerBatch` configuration. Note that the input size and output size do not have to match, unlike with most Pandas UDFs.

 All the data of a cogroup will be loaded into memory, which means if there is data skew or certain groups are too big to fit in memory you could run into OOM issues.

Spark 3.0 also introduces cogrouped map Pandas UDFs. The `applyInPandas()` function takes two `pandas.DataFrames` that share a common key and applies a function to each cogroup. The returned `pandas.DataFrames` are then combined as a single DataFrame. As with `mapInPandas()`, there is no restriction on the length of the returned `pandas.DataFrame`. Here's an example:

```
df1 = spark.createDataFrame(
    [(1201, 1, 1.0), (1201, 2, 2.0), (1202, 1, 3.0), (1202, 2, 4.0)],
    ("time", "id", "v1"))
df2 = spark.createDataFrame(
    [(1201, 1, "x"), (1201, 2, "y")], ("time", "id", "v2"))

def asof_join(left: pd.DataFrame, right: pd.DataFrame) -> pd.DataFrame:
    return pd.merge_asof(left, right, on="time", by="id")

df1.groupby("id").cogroup(
    df2.groupby("id")
).applyInPandas(asof_join, "time int, id int, v1 double, v2 string").show()
```

```
+----+---+---+---+
|time| id| v1| v2|
+----+---+---+---+
|1201|  1|1.0|  x|
|1202|  1|3.0|  x|
|1201|  2|2.0|  y|
|1202|  2|4.0|  y|
+----+---+---+---+
```

Changed Functionality

Listing all the functionality changes in Spark 3.0 would transform this book into a brick several inches thick. So, in the interest of brevity, we will mention a few notable ones here, and leave you to consult the release notes for Spark 3.0 for full details and all the nuances as soon as they are available.

Languages Supported and Deprecated

Spark 3.0 supports Python 3 and JDK 11, and Scala version 2.12 is required. All Python versions earlier than 3.6 and Java 8 are deprecated. If you use these deprecated versions you will get warning messages.

Changes to the DataFrame and Dataset APIs

In previous versions of Spark, the Dataset and DataFrame APs had deprecated the `unionAll()` method. In Spark 3.0 this has been reversed, and `unionAll()` is now an alias to the `union()` method.

Also, earlier versions of Spark's `Dataset.groupByKey()` resulted in a grouped Dataset with the key spuriously named as `value` when the key was a non-struct type (`int`, `string`, `array`, etc.). As such, aggregation results from `ds.groupByKey().count()` in the query when displayed looked, counterintuitively, like (`value`, `count`). This has been rectified to result in (`key`, `count`), which is more intuitive. For example:

```scala
// In Scala
val ds = spark.createDataset(Seq(20, 3, 3, 2, 4, 8, 1, 1, 3))
ds.show(5)
```

```
+-----+
|value|
+-----+
|   20|
|    3|
|    3|
|    2|
|    4|
+-----+
```

```
ds.groupByKey(k=> k).count.show(5)

+---+--------+
|key|count(1)|
+---+--------+
|  1|       2|
|  3|       3|
| 20|       1|
|  4|       1|
|  8|       1|
+---+--------+
only showing top 5 rows
```

However, you can preserve the old format if you prefer by setting spark.sql.leg acy.dataset.nameNonStructGroupingKeyAsValue to true.

DataFrame and SQL Explain Commands

For better readability and formatting, Spark 3.0 introduces the Data Frame.explain(*FORMAT_MODE*) capability to display different views of the plans the Catalyst optimizer generates. The *FORMAT_MODE* options include "simple" (the default), "extended", "cost", "codegen", and "formatted". Here's a simple illustration:

```
// In Scala
val strings = spark
 .read.text("/databricks-datasets/learning-spark-v2/SPARK_README.md")
val filtered = strings.filter($"value".contains("Spark"))
filtered.count()

# In Python
strings = spark
 .read.text("/databricks-datasets/learning-spark-v2/SPARK_README.md")
filtered = strings.filter(strings.value.contains("Spark"))
filtered.count()

// In Scala
filtered.explain("simple")

# In Python
filtered.explain(mode="simple")

== Physical Plan ==
*(1) Project [value#72]
+- *(1) Filter (isnotnull(value#72) AND Contains(value#72, Spark))
   +- FileScan text [value#72] Batched: false, DataFilters: [isnotnull(value#72),
Contains(value#72, Spark)], Format: Text, Location:
InMemoryFileIndex[dbfs:/databricks-datasets/learning-spark-v2/SPARK_README.md],
PartitionFilters: [], PushedFilters: [IsNotNull(value),
StringContains(value,Spark)], ReadSchema: struct<value:string>

// In Scala
filtered.explain("formatted")
```

```
# In Python
filtered.explain(mode="formatted")

== Physical Plan ==
* Project (3)
+- * Filter (2)
   +- Scan text  (1)

(1) Scan text
Output [1]: [value#72]
Batched: false
Location: InMemoryFileIndex [dbfs:/databricks-datasets/learning-spark-v2/...
PushedFilters: [IsNotNull(value), StringContains(value,Spark)]
ReadSchema: struct<value:string>

(2) Filter [codegen id : 1]
Input [1]: [value#72]
Condition : (isnotnull(value#72) AND Contains(value#72, Spark))

(3) Project [codegen id : 1]
Output [1]: [value#72]
Input [1]: [value#72]

-- In SQL
EXPLAIN FORMATTED
SELECT *
FROM tmp_spark_readme
WHERE value like "%Spark%"

== Physical Plan ==
* Project (3)
+- * Filter (2)
   +- Scan text  (1)

(1) Scan text
Output [1]: [value#2016]
Batched: false
Location: InMemoryFileIndex [dbfs:/databricks-datasets/
learning-spark-v2/SPARK_README.md]
PushedFilters: [IsNotNull(value), StringContains(value,Spark)]
ReadSchema: struct<value:string>

(2) Filter [codegen id : 1]
Input [1]: [value#2016]
Condition : (isnotnull(value#2016) AND Contains(value#2016, Spark))

(3) Project [codegen id : 1]
Output [1]: [value#2016]
Input [1]: [value#2016]
```

To see the rest of the format modes in action, you can try the notebook in the book's GitHub repo (*https://github.com/databricks/LearningSparkV2*). Also check out the migration guides (*https://spark.apache.org/docs/latest/migration-guide.html*) from Spark 2.x to Spark 3.0.

Summary

This chapter provided a cursory highlight of new features in Spark 3.0. We took the liberty of mentioning a few advanced features that are worthy of note. They operate under the hood and not at the API level. In particular, we took a look at dynamic partition pruning (DPP) and adaptive query execution (AQE), two optimizations that enhance Spark's performance at execution time. We also explored how the experimental Catalog API extends the Spark ecosystem to custom data stores for sources and sinks for both batch and streaming data, and looked at the new scheduler in Spark 3.0 that enables it to take advantage of GPUs in executors.

Complementing our discussion of the Spark UI in Chapter 7, we also showed you the new Structured Streaming tab, providing accumulated statistics on streaming jobs, additional visualizations, and detailed metrics on each query.

Python versions below 3.6 are deprecated in Spark 3.0, and Pandas UDFs have been redesigned to support Python type hints and iterators as arguments. There are Pandas UDFs that enable transforming an entire DataFrame, as well as combining two cogrouped DataFrames into a new DataFrame.

For better readability of query plans, `DataFrame.explain(`*FORMAT_MODE*`)` and `EXPLAIN` *FORMAT_MODE* in SQL display different levels and details of logical and physical plans. Additionally, SQL commands can now take join hints for Spark's entire supported family of joins.

While we were unable to enumerate all the changes in the latest version of Spark in this short chapter, we urge that you explore the release notes when Spark 3.0 is released to find out more. Also, for a quick summary of the user-facing changes and details on how to migrate to Spark 3.0, we encourage you to check out the migration guides.

As a reminder, all the code in this book has been tested on Spark 3.0.0-preview2 and should work with Spark 3.0 when it is officially released. We hope you've enjoyed reading this book and learned from this journey with us. We thank you for your attention!

Index

B

bagging, 313
barrier execution mode, 351
batch deployment, 332
Beeline, querying with, 119
BHJ (broadcast hash join), 188, 349
big data, 1
Big table, 1
bin directory, 21, 21
binary files, as a data source for DataFrames
and SQL tables, 110
bootstrapping samples, 313
broadcast variables, 188, 344
bucketBy() method, 96
built-in data sources, 83-112, 94
built-in functions, 139-141, 239
bytecode, 7, 23, 25

C

cache(), 183-187
caching, 93, 183-187
cardinality() function, 139
case class, 71, 158
CASE statement, 152
Cassandra, 89, 137, 231
Catalog API, 93, 349-351
Catalyst optimizer, xvi, 16, 77-82, 170
CBO (cost-based optimizer), 81
CDC (change-data-capture), 271
checkpointing, 217, 262
classification, 286-287, 292, 304
clause conditions, 281
client mode, 12
close() method, 233
cluster managers, 10, 12, 176, 178
cluster resource provisioning, 262
clustering, 286, 288, 302
code examples, using, xviii
Code generation phase (Spark SQL), 81
codegen, enabling in Spark SQL, 189
cogroup(), 356
collect() method, 67
collect_list(), 138
collect_set(), 239
Column object, 54
columns
    adding, 63, 152
    dropping, 63, 152
    in DataFrames, 54

random feature selection by, 313
renaming, 63, 153
comma-separated value files (CSV files), 102
community adoption/expansion, of Spark, 16
Complete mode (Structured Streaming), 212,
    215, 245
complex data types, 49, 139-141
compression property, 101, 103, 106
compute function, 44
concat() function, 139
conf.spark-defaults.conf file, 173
configurations
    setting, 173-176
    used in this book, xviii
    viewing, 173-176
configuring Spark, with Delta Lake, 274
consistency, of data lakes, 270
continuous applications, 15
Continuous mode (Structured Streaming), 217
Continuous Streaming model, 8
Continuous trigger mode, 219
correlation() method, 68
costs
    mitigating, 170
    of databases, 267
    of latency, 209
count(), 29, 66, 183, 215
countDistinct(), 239
Counting M&Ms example, 35-39
covariance() method, 68
CrossValidator, 317
CSV files, as a data source for DataFrames and
    SQL tables, 102
cube() function, 235
cubed() function, 116
customStateUpdateFunction(), 254

D

DAG (directed acyclic graph), 4, 27
data
    accommodating changing, 279
    auditing changes with operation history,
        282
    deduplicating, 281
    diversity of formats for storage solutions,
        265
    governance of, as a feature of lakehouses,
        271
    growth in size of, 267

F

fault tolerance, 2, 9, 15, 185, 209, 222
fault-tolerant state management, 236
file formats
    about, 76
    CSV files, 102
    data lakes and, 269
    support for diversity of, 269
files
    about, 21
    reading from, 226
    Structured Streaming and, 226
    writing to, 227
filesystems, 89, 269
filter() method, 28, 29, 61, 72, 73, 143, 157, 162, 170, 215, 235
filtering, DataFrames and, 61
fit() method, 296
fitting, 295
flatMap() method, 157, 170, 215, 235
flatMapGroupsWithState(), 253, 256, 261
flatten() function, 139
fmin(), 339
foreach() method, 216, 230, 233-234
foreachBatch() method, 216, 230, 281
format() method, 94, 96, 100
frequentItems() method, 68
Friedman, Jerome, The Elements of Statistical Learning, 309
from_json() function, 138
functional programming, higher-order functions and, 162-167
functionality, changed, 357-360

G

garbage collection, 167, 178, 199
GDPR (General Data Protection Regulation), 280
generalization, with flatMapGroupsWithState(), 261
generic rows, 69
getter methods, 70
get_json_object() function, 138
GFS (Google File System), 1
Ghemawat, Sanjay, The Google File System, 268
global aggregations, 238
global temporary views, 92
Gobioff, Howard, The Google File System, 268
Google, 1

The Google File System (Ghemawat, Gobioff, and Leung), 268
GraphFrames, 9
graphical user interface, 31
GraphX library, 6, 9
GridSearchCV, 339
GROUP BY statement, 138
groupBy() method, 30, 66, 73, 157, 182, 187, 244, 337
groupByKey(), 254, 256
grouped aggregate Pandas UDFs, 116
grouped aggregations, 238
grouped map Pandas UDFs, 116

H

Hadoop, 2, 268
Hadoop YARN, 12
Hastie, Trevor, The Elements of Statistical Learning, 309
HBase, 5
HDFS (Hadoop Distributed File System), 2, 268
high-level structured APIs, 25
higher-order functions, 138-144, 162-167
Hive, 89, 113-155
Hive ORC SerDe (serialization and deserialization) tables, 107
HiveContext object, 11
HiveServer2, 120
Hyperopt, 339
hyperparameter configurations, 317
hyperparameter tuning
    about, 307
    distributed, 337-340
    k-fold cross-validation, 316-319
    optimizing pipelines, 320-321
    tree-based models, 307-316

I

id column (StreamingQuery), 224
ignoreExtension property, 106
images, as a data source for DataFrames and SQL tables, 108
incremental execution, 234
incrementalization, 211
inferSchema property, 103
inner joins, 248-252
input and output sources
    defining, 213
    file formats

M

machine learning (ML)
about, 286
building models using estimators, 295
creating pipelines, 296-302
creating test data sets, 291-293
creating training data sets, 291-293
data ingestion, 290
designing pipelines, 289-307
evaluating models, 302-306
exploration, 290
hyperparameter tuning, 307-321
linear regression, 294
loading models, 306
reasons for using Spark, 289
saving models, 306
supervised, 286
unsupervised, 288
with MLlib, 285-321
machine learning engineers, xv
managed stateful transformations, 237
managed tables, 89
map functions, 139
map() method, 73, 138, 157, 162, 163, 170, 215, 235
map-side-only join, 188
mapGroupsWithState(), 253, 256, 261
mapInPandas() method, 336, 356
mapPartitions(), 119
map_concat() function, 139
map_form_arrays() function, 139
map_from_entries() function, 139
Mastering Spark with R (Luraschi, Kuo, and Ruiz), xvi, 285
Matrix object, 292
Maven, 133, 134
max() method, 67
mean() method, 239
memory management, for Datasets and Data-Frames, 167
MEMORY_AND_DISK storage level, 184
MEMORY_AND_DISK_SER storage level, 184
MEMORY_ONLY storage level, 184
MEMORY_ONLY_SER storage level, 184
merge(), upserting change data to tables using, 281
Mesos (see Apache Mesos)
metadata, 93, 326
metrics, 224, 326

micro-batch architecture, Spark Streaming, 208
min() method, 67
mitigating costs, 170
MLeap, 334
MLflow, 323, 324-330
MLflow Model Registry, 332
MLflow Models, 332
MLflow Projects, 330
MLflow Tracking, 325-330
MLlib library
about, xv, 6, 7
(see also machine learning (ML))
machine learning (ML) with, 285-321
model deployment options with, 330
model.transform(), 332
models
about, 326
building using estimators, 295
evaluating, 302-306
loading, 306
managing, 323-330
saving, 306
tree-based, 307-316
models component (MLflow), 324
modifications, 151-155
modularity, of Spark, 5
MongoDB, 137
month() function, 65
MR (MapReduce), 1
MS SQL Server, 136
multiline property, 101, 103
multitenant environment, 178
MySQL database, 86, 133

N

Naive Bayes algorithm, 287
narrow dependencies, 30
Netflix, 272
non-MLlib models, leveraging Spark for, 336-340
non-SQL based analytics, for databases, 268
non-time based streaming aggregations, 238
null checking, 115
numInputRows column (StreamingQuery), 224
numPartitions property, 130
NumPy, 15

O

objects, 69

R

R library, 21
R2, 302-306
R2D3, 309
random forests, 313-316
randomSplit(), 293
ranking functions, 149
RDBSs (relational database management systems), 1
RDD (Resilient Distributed Dataset), 5, 16, 43, 75
rdd.getNumPartitions(), 13
read(), 29
reading
    Avro files into DataFrames, 104
    Avro files into Spark SQL tables, 105
    binary files into DataFrames, 110
    CSV files into DataFrames, 102
    CSV files into Spark SQL tables, 102
    from data lakes, 269
    from databases, 267
    from files, 226
    from Kafka, 228
    image files into DataFrames, 108
    JSON files into DataFrames, 100
    JSON files into Spark SQL tables, 100
    ORC files into DataFrames, 107
    ORC files into Spark SQL tables, 107
    Parquet files into DataFrames, 97
    Parquet files into SQL tables, 97
    tables into DataFrames, 93
README.md file, 21
real-time inference, export patterns for, 334
receivers, 169
record Name property, 106
record-at-a-time processing model, 207
recordNamespace property, 106
redesigning Pandas UDFs, 354
reduce() function, 144, 162
reduceByKey(), 187
registry component (MLflow), 324
regression
    decision trees, 308-313
    linear, 294
    logistic, 287
    random forests, 313-316
rename() method, 153
renaming columns, 63, 153
REST API, 324

reverse() function, 139
RFormula, 300
RISELab, 3
RMSE (root-mean-square error), 302
rollup(), 235
root, of decision trees, 308
Row objects, 57
rows
    generic, 69
    in DataFrames, 57
    random feature selection by, 313
Ruiz, Edgar, Mastering Spark with R, xvi, 285
runID column (StreamingQuery), 224
running Spark SQL queries, 120
runs, 325
runtime architecture (Spark), 11, 74, 96, 170, 203, 234, 345

S

sample data, 160, 162-167
sampleBy() method, 68
save() method, 96
saveAsTable() method, 96
saving models, 306
sbt (Scala build tool), 40
Scala
    building standalone applications in, 40
    case classes in, 71
    columns and, 55
    single API for, 157-160
    using, 22-25
Scala shell, 23, 274
scalability
    of databases, 268
    of storage solutions, 265
    Spark, 177-182
scalar Pandas UDFs, 116
SCD (slowly changing dimension), 271
schedulers, 351
schema enforcement/governance, 271, 278-279
schema() method, 94
SchemaRDDs, 43
schemas, 50-54, 279
scikit-learn, 289, 310, 312, 336, 339
second-generation Tungsten engine, 167
select() method, 28, 61, 73, 162, 215, 235
selectExpr() function, 138
semantic guarantees, with watermarks, 245
sep property, 103

About the Authors

Jules S. Damji is a senior developer advocate at Databricks and an MLflow contributor. He is a hands-on developer with over 20 years of experience and has worked as a software engineer at leading companies such as Sun Microsystems, Netscape, @Home, Loudcloud/Opsware, Verisign, ProQuest, and Hortonworks, building large-scale distributed systems. He holds a B.Sc. and an M.Sc. in computer science and an MA in political advocacy and communication from Oregon State University, Cal State, and Johns Hopkins University, respectively.

Brooke Wenig is a machine learning practice lead at Databricks. She leads a team of data scientists who develop large-scale machine learning pipelines for customers, as well as teaching courses on distributed machine learning best practices. Previously, she was a principal data science consultant at Databricks. She holds an M.S. in computer science from UCLA with a focus on distributed machine learning.

Tathagata Das is a staff software engineer at Databricks, an Apache Spark committer, and a member of the Apache Spark Project Management Committee (PMC). He is one of the original developers of Apache Spark, the lead developer of Spark Streaming (DStreams), and is currently one of the core developers of Structured Streaming and Delta Lake. Tathagata holds an M.S. in computer science from UC Berkeley.

Denny Lee is a staff developer advocate at Databricks who has been working with Apache Spark since 0.6. He is a hands-on distributed systems and data sciences engineer with extensive experience developing internet-scale infrastructure, data platforms, and predictive analytics systems for both on-premises and cloud environments. He also has an M.S. in biomedical informatics from Oregon Health and Sciences University and has architected and implemented powerful data solutions for enterprise healthcare customers.

Colophon

The animal on the cover of *Learning Spark*, Second Edition, is the small-spotted catshark (*Scyliorhinus canicula*), an abundant species in the shallow waters of the Mediterranean Sea and in the Atlantic, off the coast of Europe and northern Africa. It is a small, slender shark with a blunt head, oval eyes, and a rounded snout. The dorsal surface is grayish-brown and patterned with many small dark and sometimes lighter spots. Like other sharks, its skin texture is formed of "dermal denticles," tiny "teeth" that grow all in one direction (like fish scales), forming a surface that's both hydrodynamic as well as resistant to injuries and parasites.

This night-feeding shark grows to about 3 feet long, weighs an average of 3 pounds at maturity, and in the wild can live up to 12 years. It feeds mostly on mollusks, crustaceans, cephalopods, and polychaete worms, though it also eats other fish. This species exhibits some social behaviors, especially when young, and a 2014 study conducted by the University of Exeter found that individuals displayed differing social personalities. Across changes in habitat, some sharks preferred staying in conspicuous groups, while others remained alone, camouflaged at the bottom of the habitat. These socialization behaviors also reflect a variability in strategies for safety, either through numbers or via camouflage.

This catshark is oviparous (egg-laying), and females deposit 18-20 small egg cases each year. These hard-shelled cases have tendrils that catch on seaweed at the ocean floor; each case contains one young shark. The young hatch after about nine months.

Because the small-spotted catshark is undesirable to commercial fisheries, populations are currently stable and the species is listed by the IUCN as being of Least Concern. Many of the animals on O'Reilly covers are endangered; all of them are important to the world.

The cover illustration is by Karen Montgomery, based on a black and white engraving from J. G. Wood's *Animate Creation* (1885). The cover fonts are Gilroy Semibold and Guardian Sans. The text font is Adobe Minion Pro; the heading font is Adobe Myriad Condensed; and the code font is Dalton Maag's Ubuntu Mono.

Printed in the USA
CPSIA information can be obtained
at www.ICGtesting.com
JSHW052151010124
54629JS00009B/136

9 781492 050049